Zinc in Human Nutrition

Author

Ananda S. Prasad

Professor of Medicine
Director, Division of Hematology
Department of Medicine
Wayne State University School of Medicine
Detroit, Michigan

CRC PRESS, INC.
Boca Raton, Florida 33431

Library of Congress Cataloging in Publication Data

Prasad, Ananda Shiva.
    Zinc in human nutrition.

    Bibliography: p.
    Includes index.

    1. Zinc deficiency diseases. 2. Zinc metabolism.
3. Zinc in the body. I. Title.
RC627.Z5P7   616.3'99   79-15272
ISBN 0-8493-0145-9

This book originally appeared as an article in Volume 8 Issue 1 of *CRC Critical Reviews in Clinical Laboratory Sciences,* a journal published by CRC Press, Inc. We would like to acknowledge the editorial assistance received by the journal's editors, Dr. John Batsakis, University of Michigan Medical School, and Dr. John Savory, University of Virginia Medical Center. The referees for this article were Professor James A. Halstead, Albany Medical College, and Dr. James C. Smith, Jr., Veterans Administration Hospital, Washington, D.C.

International Standard Book Number 0-8493-0145-9

Library of Congress Card Number 79-15272
Printed in the United States

PREFACE

Research in the field of trace elements and its application in nutrition is experiencing an exciting era of development. Although zinc has been known to be essential for plants for the past 100 years, its essentiality for the growth of animals was reported in 1934 and for man in 1963. Growth retardation, hypogonadism in the males, skin changes, poor appetite, and mental lethargy are some of the clinical features related to a deficiency of zinc in man.

Carbonic anhydrase was recognized to be the first zinc metalloenzyme in the late 1930s. However, by now there may be as many as 70 metalloenzymes that seem to require zinc for their activities.

Several enzymes required for nucleic acid metabolism have recently been shown to be zinc dependent. These include ribonucleic acid (RNA) and deoxyribonucleic acid (DNA) polymerases, reverse transcriptase, and deoxythymidine kinase. In biological experiments, it has been shown that the activity of deoxythymidine kinase is decreased in rapidly regenerating connective tissue as early as 6 days after the animals are placed on zinc-deficient diet. This is perhaps one of the earliest metabolic defects known to occur as a result of nutritional zinc deficiency in animals. These observations underscore the basic importance of zinc for cell division and protein synthesis.

For a long time, zinc deficiency in man was considered unlikely in view of the mineral's ubiquitous nature. It appears now that a nutritional zinc deficiency may be prevalent in inhabitants of many developing countries subsisting on a high cereal protein intake. The adverse effect of phytate on zinc availability and its high content in cereal proteins has been understood only recently, thus demonstrating that a deficiency of zinc in man may occur under most practical conditions. Recent reports indicate that a marginal zinc deficiency may indeed be widespread even in developed countries such as the U.S., probably related to self-imposed dietary restrictions, use of alcohol and cereal proteins, and increasing use of refined food materials which decrease the overall intake of trace elements.

The most exciting development with respect to clinical medicine appears to be the recognition that zinc may have a vital role in the management of certain diseased states. Zinc therapy appears to be a life saving measure in acrodermatitis enteropathica. The role of zinc in the management of growth and gonadal problems in malabsorption syndrome is now becoming clear. Zinc appears to be beneficial for wound healing in zinc-deficient subjects. Zinc deficiency appears to be a prominent feature of many cases of sickle cell anemia resulting in delayed growth and hypogonadism. It has now been shown that zinc may have a pharmacological antisickling effect possibly due to a newly discovered anticalcium effect in the red cell membrane. Recently, zinc has been shown to be beneficial in the management of acute inflammatory conditions associated with rheumatoid arthritis, probably related to its effects on the cell membranes. It is obvious that an awareness of the biochemical roles of zinc may find its application in management of other diseased conditions in the future.

In man, the role of zinc immunity and infections, and the effects of its deficiency on pregnancy and behavior need further investigations. Requirement of zinc in man under different dietary conditions and simple diagnostic tests to recognize marginal deficiency of zinc need to be established. In view of the known interactions of zinc with cadmium, copper, and lead, one may postulate a therapeutic role of zinc in toxicities due to these elements. In the future, these studies must be conducted with great care and caution, and an indiscriminate use of zinc by physicians must be deplored. Carefully planned and critical studies with respect to zinc metabolism in man, however, are likely to yield very valuable information in the future, which may have important implications in the management of other diseases in man.

I hope that this critical review on zinc may provide stimulus for further research. Physicians, biochemists, and nutritionists should find this review interesting in view of many new developments in this area.

THE AUTHOR

Dr. Ananda S. Prasad is Professor of Medicine and Director, Division of Hematology at Wayne State University, School of Medicine, Detroit, Michigan.

Dr. Prasad received his M.B., B.S. degree from Patna University, India and Ph.D. degree from the University of Minnesota. He is the author of numerous scientific articles and five books related to trace elements. He is a member of several prestigious scientific organizations both in the U.S. and abroad.

His awards include the Joseph Goldberger Award in Clinical Nutrition (AMA) in 1975, and the American College of Nutrition Award in 1976. He is currently editor of the American Journal of Hematology.

TABLE OF CONTENTS

Introduction 1
  Nutritional Deficiency of Zinc in Man 2
  Zinc Deficiency in Preadolescent School Children in Iran 11
  Zinc Deficiency Associated with Generalized Malnutrition 12
  Secondary Zinc Deficiency in Children 12
  Nutritional Zinc Deficiency in Children in the United States 13
  Zinc Nutrition and Deficiency in Infants 14
    Low Levels of Zinc in the Plasma and Hair of U.S. Infants 14
    Symptomatic Zinc Deficiency in Infants 15

Diagnosis of Zinc Deficiency in Man 16

Conditioned Deficiency of Zinc in Man 17
  Nutritional Factors 17
  Liver Disease 17
  Gastrointestinal Disorders 19
  Hyperzincuria and Renal Disease 20
  Neoplastic Disease 21
  Burns and Skin Disorders 21
  Impaired Wound Healing in Chronically Diseased Subjects 21
  Parasitic Infestations 22
  Iatrogenic Cause 23
  Diabetes 23
  Collagen Diseases 23
  Pregnancy and Oral Contraceptives 24
  Genetic Disorders 24
    Sickle Cell Disease 24
    Acrodermatitis Enteropathica 29
    Miscellaneous Genetic Disorders 29
    Zinc Therapy in Rheumatoid Arthritis 30

Metabolic Aspects of Zinc in Human Nutrition 30
  Distribution in the Body 30
  Zinc in Plasma and Red Cells 30
  Absorption of Zinc 32
  Availability of Zinc 33
  Geophagia 34
  Chelating Agents 34
  Intake and Excretion of Zinc in Man 35

Biochemistry and Physiology of Zinc 40
  Zinc Enzymes in Zinc Deficiency 41
  Alkaline Phosphatase in Intestinal and Other Tissues 43
  Pancreatic Carboxypeptidases 43
  Carbonic Anhydrase 44
  Alcohol Dehydrogenase 45
  Glutamic Dehydrogenase 45
  Lactic and Malic Dehydrogenase 45
  Aldolase, NADH Diaphorase, and Pyridoxal Phosphokinase 49

Reduced Enzyme Activities and Symptoms Due to Zinc Deficiency 50
Total Protein in Zinc Deficiency 50
Changes in Nucleic Acid Metabolism 51
Tissue Levels of RNA and DNA 51
Catabolism and Synthesis of RNA and DNA 51
Polynucleotide Conformation 53
Zinc and Hormones 53
    Glucose Tolerance 53
    Insulin 56
    Growth and Sex Hormones 57
Zinc in Collagen Metabolism 57
Zinc and Cystine Metabolism 63
Effect of Zinc on Cell Membrane 63
Zinc as an Integral Part of Plasma Membrane of Macrophages 65
Lymphocytes and Zinc 65
Effects on Red Cells 66
Zinc as a Viral Inhibitor 67
Zinc and Metallothionein 67

Toxicity 67

Acknowledgment 68

References 68

Index 81

ZINC IN HUMAN NUTRITION

Author: **Ananda S. Prasad**
            Division of Hematology
            Department of Medicine
            Wayne State University School of Medicine
            Detroit, Michigan

Referees: James A. Halsted
            Albany Medical College
            Albany, New York

            James C. Smith, Jr.
            Trace Element Research Laboratory
            Veterans Administration Hospital and
                Department of Biochemistry
            Medical School
            George Washington University
            Washington, D.C.

INTRODUCTION

In 1869 Raulin[1] first showed that zinc was essential for the growth of *Aspergillus niger*. This was confirmed almost 40 years later by Bertrand and Javillier.[2] Its essentiality for higher forms of plant life was established in 1926.[3,4]

In 1934 Todd, Elvehjem, and Hart[5] reported that zinc was necessary for the growth and well being of the rat. Previous attempts by Bertrand and Benson,[6] McHargue,[7] and Hubbell and Mendel[8] to demonstrate the essentiality of zinc in animals were unsuccessful because the purified diets used in their experiments were deficient in other essential nutrients. Tucker and Salmon[9] reported that zinc cures and prevents parakeratosis in swine. In 1958, O'Dell and Savage[10] and O'Dell, Newberne, and Savage[11] showed that zinc was required for bodily functions and growth in birds. Zinc deficiency occurs in suckling mice that are deprived of colostrum.[12] The manifestations of zinc deficiency consisted of retarded growth and ossification, alopecia, thickening and hyperkeratinization of the epidermis, clubbed digits, deformed nails, and moderate congestion in certain viscera. Experimental zinc deficiency in calves has been produced.[13,14] The main features include retarded growth, testicular atrophy, and hyperkeratosis. Deficiency of zinc in the diet of breeding hens results in lowered hatchability, gross embryonic anomalies characterized by impaired skeletal development, and varying degrees of weakness in chicks that hatch.[15] Zinc deficiency in dogs has been produced by feeding them a diet low in zinc and high in calcium.[16] Signs of zinc deficiency in dogs include retardation of growth, emaciation, emesis, conjunctivitis, keratitis, general debility, and skin lesions on the abdomen and extremities. Zinc deficiency in young Japanese quail has been produced by feeding a zinc low-purified diet containing soybean protein.[17] Slow growth, abnormal feathering, labored respiration, incoordinate gait, and low content of zinc in the liver and tibias were noted. Zinc deficiency has now been reported in approximately 15 species including man.

Although essentiality of zinc for animals was recognized, its ubiquity made it seem unlikely that alterations in zinc metabolism could lead to significant problems in human nutrition or clinical medicine. This attitude has now changed. Zinc deficiency in man was suspected to occur for the first time in 1961[18] in Iranian males and was established following detailed studies in Egypt in 1963.[19,23]

In the fall of 1958, Dr. James A. Halsted brought to my attention a 21-year-old patient at Saadi Hospital, Shiraz, Iran, who looked like a 10-year-old boy. In addition to dwarfism and severe anemia, he had hypogonadism, hepatosplenomegaly, rough and dry skin, mental lethargy, and geophagia. The nutritional history was interesting in that this patient ate only bread from wheat flour and the intake of animal protein was negligible. He consumed nearly 1 lb of clay daily. Later we discovered that the habit of geophagia (clay-eating) is not uncommon in the villages around Shiraz. Through further investigations, it was obvious that this patient had severe iron deficiency. There was no evidence of blood loss. Hookworm and schistosomiasis infestations are not seen in that part of Iran. Shortly thereafter, ten similar cases were investigated in detail. The probable factors responsible for anemia in these patients were

1. The total amount of available iron in the diet was insufficient.

2. Excessive sunburn and sweating probably caused greater iron loss from the skin than would occur in a temperate climate.

3. Geophagia may have further decreased iron absorption, as observed by Minnich et al.[20]

In every case the anemia was completely corrected by administration of oral iron.[18]

Lemann previously observed this clinical syndrome in 1910 in the United States.[21] However, it was not related to a nutritional deficiency. Similar cases from Turkey have been reported,[22] but detailed descriptions were not given, and the authors considered a genetic defect to be a possible explanation for certain aspects of the clinical picture. Prasad, Halsted, and Nadimi published a detailed clinical report from Iran in 1961.[18] Although no data were available to document zinc deficiency in the patients, this possibility was considered. Subsequently, our studies in Egypt established that such patients were zinc deficient,[19,23,35] thus demonstrating for the first time that deficiency of zinc may occur in man.

Nutritional Deficiency of Zinc in Man

In 1961 a group of 11 Iranian adult males were reported to show the following clinical features:[18] severe iron deficiency anemia, hepatosplenomegaly, short stature, and marked hypogonadism[18] (Figure 1). Their diet consisted almost exclusively of bread made from wheat flour (the intake of animal protein was negligible). They all gave a history of geophagia. There was no evidence of blood loss or hookworm infestation. The anemia promptly responded to oral administration of pharmaceutical iron. Following therapy with orally administered pharmaceutical ferrous sulfate (1 g daily) and a good hospital diet, the anemia was corrected, their hepatosplenomegaly improved, they grew pubic hair, and their genitalia size increased. Liver function tests were not remarkable, except for the serum alkaline phosphatase which increased following treatment (Figure 2).

It was difficult to explain all of the clinical features solely on the basis of tissue iron deficiency. Tissue effects of iron deficiency in animal and human subjects have been adequately described.[24] Iron-deficient, New Hampshire chicks have white feathers rather than their normally reddish-brown plumage. The iron-deficient rat loses the deep orange pigment of its incisors. A decrease in gastric acid secretion and an increase in the size of the cecum have been noted in rats given an iron-poor diet. The ability to absorb vitamin B_{12} by iron-deficient rats, perhaps caused by defective intrinsic factor secretion, has also been described. Rats and elephants continue to grow on an iron-deficient diet, but develop marked anemia during the period of rapid growth.[25]

In humans, iron deficiency causes changes in the mucosa of the alimentary tract.[24,] The oral mucosal epithelium is abnormally thin in some patients. Mitoses in the prickle cell layer were more frequent in iron-deficient subjects than in normal ones. Melanin deposition is decreased, and subepithelial inflammation is increased. Changes in the esophagus of iron-deficient patients give rise to "sideropenic dysphagia." Constrictions or webs of

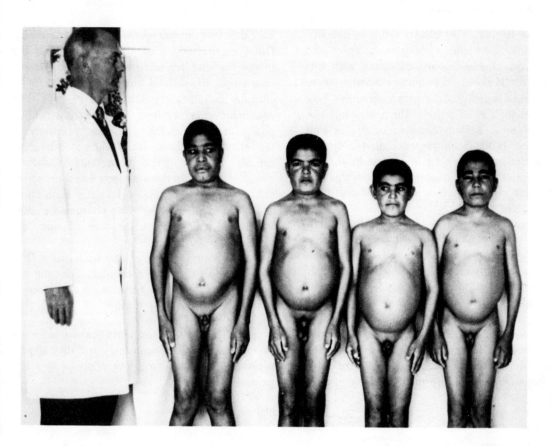

FIGURE 1. Four dwarfs from Iran. From left to right: age 21, height 4 ft, 11½ in.; age 18, height 4 ft, 9 in.; age 18, height 4 ft, 7 in.; age 21, height 4 ft, 7 in. Staff physician (left) is 6 ft in height. (From Prasad, A. S., Halsted, J. A., and Nadimi, M., *Am. J. Med.,* 31, 352, 1961. With permission.)

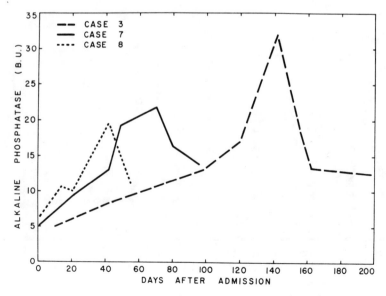

FIGURE 2. Changes in serum alkaline phosphatase associated with hospitalization and a well-balanced diet in cases of Iranian drawfs. (From Prasad, A. S., Halsted, J. A., and Nadimi, M., *Am. J. Med.,* 31, 352, 1961. With permission.)

the hypopharynx or esophagus result in inability to swallow solid food.[27] Gastric atrophy and achlorhydria are commonly associated with this disorder.[28] However, it is unclear whether these are the results of iron deficiency or whether they are of etiologic significance in the development of hyposideremia. Koilonychia also occurs in iron deficiency. It has been suggested that a dietary cystine deficiency may be important in the development of this abnormality.[29] Koilonychia and esophageal changes have also been described in zinc-deficient animals.[12,30]

It is unlikely that iron deficiency alone could account for all the clinical features noted in the patients described above. The possibility that zinc deficiency may have been present was considered. As noted earlier, zinc deficiency is known to produce retardation of growth and testicular atrophy in animals. Inasmuch as heavy metals may form insoluble complexes with phosphates, one may speculate that some factors responsible for decreased availability of iron in these patients with geophagia may also govern the availability of zinc.

Zinc deficiency in rats results in testicular atrophy. Other manifestations of zinc deficiency in the mouse, rat and pig include lack of growth and retardation of skeletal maturation. Changes in alkaline phosphatase (widely held to be a zinc-containing enzyme) have also been observed in pigs with zinc deficiency. Increasing activity of this enzyme has been noted when zinc-deficient animals received increased amounts of dietary zinc.[31,32] Thus, in these subjects dwarfism, testicular atrophy, retardation or skeletal maturation, and changes in serum alkaline phosphatase could have been explained on the basis of zinc deficiency.

Subsequently, in Egypt, similar patients (Figures 3 and 4) were encountered in the villages.[19,23] The clinical features were remarkably similar, except for the following:

1. The Iranian patients exhibited more pronounced hepatosplenomegaly, and they all gave a history of geophagia; however, none had any parasitic infestations.

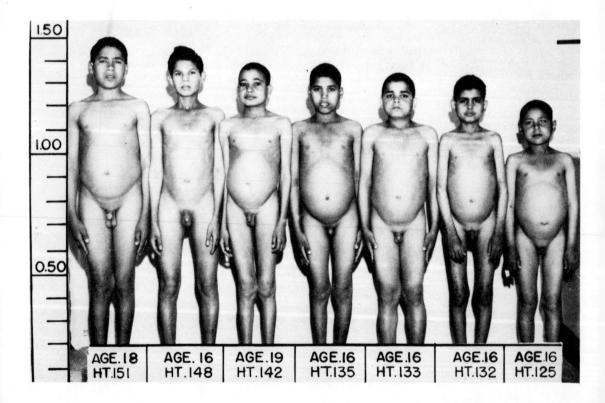

FIGURE 3.   Seven dwarfs from delta villages near Cairo, Egypt. Height is shown in centimeters. (From Prasad, A. S., Miale, A., Farid, Z., Schulert, A., and Sandstead, H. H., *J. Lab. Clin. Med.*, 61, 531, 1963. With permission.)

FIGURE 4. Sixteen-year-old boys. The one on the left measures 138 cm in height and 36 kg in weight; the one on the right measures 145 cm in height and 41 kg in weight. (From Prasad, A. S., Schulert, A. R., Miale, A., Farid, Z., and Sandstead, H. H., *Am. J. Clin. Nutr.,* 12, 437, 1963. With permission.)

2. The majority of Egyptian patients had both schistosomiasis and hookworm infestations, and none gave a history of geophagia.

The Egyptian patients were studied in detail. Their dietary history was similar to that of the Iranians. The intake of protein was negligible, and their diet consisted mainly of bread and beans (*Vicia fava*). These subjects were found to have a zinc deficiency. This conclusion was based on the following:

1. The zinc concentrations in plasma, red cells, and hair were decreased.

2. Radioactive zinc-65 studies revealed that the plasma zinc turnover rate was greater, the 24-hr exchangeable pool was smaller, and the excretion of zinc-65 in stool and urine was less in the patients than in the control subjects.

These data are summarized in Tables 1 and 2. Zinc deficiency in humans, in the absence of advanced cirrhosis of the liver, had not been described before. Liver function tests and biopsy failed to reveal evidence of cirrhosis of the liver in these subjects.[33] Furthermore, in contrast to cirrhotic patients who excrete abnormally high quantities of zinc in urine,[34] our patients excreted less stable zinc in urine, as compared to control subjects.[33] These results are presented in Tables 3 and 4. Detailed examination of these patients ruled out other chronic debilitating diseases which might affect the serum zinc levels.

Investigations for deficiency of other metals were also conducted. Serum iron was decreased, unsaturated iron-binding capacity was increased, serum copper was slightly increased, and serum magnesium was normal. Hair analyzed for manganese, cobalt, molybdenum, and other elements revealed no significant decrease as compared to the normal subjects (Tables 5 and 6).

Investigations for vitamin deficiency in these patients were also nonrevealing. Serum B_{12}, ascorbic acid, vitamin A, and carotene levels were not abnormally low. Formiminoglutamic acid excretion following histidine loading and xanthurenic acid excretion following tryptophane loading were also normal, thus indicating that folic acid and vitamin B_6 deficiencies were not implicated.[19]

Changes in serum alkaline phosphatase (Figure 5), similar to those observed in Iranian patients, were seen in the Egyptian cases.[19,23] In serum, alkaline phosphatase activity increases with administration of growth hormone;[36] however, many other factors also influence its activity. Thus, the increasing activity of this enzyme may be indicative of increased production of the growth hormone following treatment. In Iran, patients with this syndrome treated with iron and a good diet developed secondary sexual characteristics, gained in weight and height, showed an increase in the size of the liver and spleen, and exhibited correction of anemia after adequate treatment.[18] Ordinary pharmaceutical preparations of iron may contain appreciable quantities of zinc as a con-

TABLE 1

Zinc Content of Plasma, Erythrocytes, and Hair

	Plasma[a] (μg %)	RBC[a] (μg/ml)	Hair[a] (μg/g)
Normals	102 ± 13 (19)[b]	12.5 ± 1.2 (15)	99 ± 9 (10)
Dwarfs	67 ± 11 (17)	9.7 ± 1.1 (14)	65 ± 16 (10)

Note: ±, standard deviation.

[a]The differences between normals and dwarfs are statistically significant (p < 0.01).

[b]Numbers in parentheses indicate the number of subjects included for each determination.

From Prasad, A. S., in *Zinc Metabolism,* Prasad, A. S., Ed., Charles C Thomas, Springfield, Ill., 1966, 268. With permission.

TABLE 2

Summary of Zinc-65 Studies

	Turnover rate[a] (mg/kg/day)	24-hr exchangeable pool[a] (mg/kg)	Urinary excretion[a] in % dose administered in 13 days	Excretion in stool[b] in % dose administered in 100 g of stool
Normals	1.00 ± 0.09 (9)[c]	7.0 ± 1.6 (8)	2.8 ± 0.56 (7)	0.66 ± 0.19 (7)
Dwarfs	1.50 ± 0.29 (10)	4.6 ± 1.2 (8)	1.6 ± 0.68 (7)	0.42 ± 0.13 (7)

Note: ±, standard deviation.

[a]The differences between normals and dwarfs are statistically significant (p < 0.01).
[b]p < 0.05.
[c]Numbers in parentheses indicate the number of subjects included for each determination.

From Prasad, A. S., in *Zinc Metabolism,* Prasad, A. S., Ed., Charles C Thomas, Springfield, Ill., 1966, 268. With permission.

taminant and thus could supply enough zinc to institute recovery from a deficient state.

It is a common belief among medical practitioners in Iran that severe retardation of growth and sexual hypofunction, as noted above, are the results of visceral leishmaniasis and geophagia. In our detailed investigations, no evidence of visceral leishmaniasis was found. The role of geophagia is not entirely clear at the present time; however, it is believed that the excess amount of phosphate in the clay may prevent absorption of both dietary iron and zinc. The predominantly wheat diet in the Middle East, which has a high content of phytate and fiber, may also reduce the availability of zinc.[37]

In Egypt, the cause of dwarfism was commonly considered to be schistosomiasis.[39] The liver dysfunction that results from schistosomiasis is regarded as the cause of hypogonadism. Chinese investigators have also suggested that schistosomiasis and associated liver disease may cause dwarfism and hypogonadism.[39]

Since the Iranian patients exhibited dwarfism, but did not have schistosomiasis or hookworm infections, the question arose as to whether or not schistosomiasis was the fundamental cause of dwarfism among the Egyptian patients; an investigation was undertaken to answer this question. It was known that there was no schistosomiasis or hookworm infection in the villages of Kharga, a desert oasis that is 500 km southwest of Cairo.[40] Culturally and nutritionally speaking, however, the

TABLE 3

Liver Function Studies in Patients with Anemia, Dwarfism, and Hypogonadism[a]

No.	BSP[b] retention after 45 min (%)	Total serum bilirubin (mg %)	Serum alkaline phosphatase	SGOT units	SGPT units	LDH	Serum total protein (g %)	Albumin (g %)	Thymol turbidity units	Cephalin flocculation 24 hr and 48 hr	Liver biopsy
1	0.3	0.5	8.4 BU	14	7	430	7.9	4.1	4.5	0 1+	Minimal histologic changes
2	0.3	0.4	1.45 SU[b]	18	5	282	6.4	4.3	2.8	1+, 2+	Minimal histologic changes
3	4.7	0.33	2.35 SU[b]	16	9	473	7.0	4.1	3.0	2+, 2+	Minimal histologic changes
4	0.6	0.33	4.55 SU[b]	33	8	375	7.7	4.6	5.0	1+, 2+	Minimal histologic changes
5	2.6	0.4	10.9 BU	32	18	240	7.5	3.9	4.0	2+, 2+	Minimal focal necrosis and periportal infiltrate
6	1.5	0.33	12.8 BU	26	16	325	7.9	4.4	7.7	3+, 4+	Mild focal necrosis and periportal infiltrate
7	0.5	0.33	2.0 SU[b]	33	20	323	8.3	3.8	3.5	2+, 3+	Moderate focal necrosis and periportal infiltrate. Granulomas with schistosoma ova
8	0.5	0.25	4.1 SU[b]	29	18	295	8.6	3.9	7.0	2+, 2+	Minimal focal necrosis and periportal infiltrate

Note: BSP, bromosulfophthalein; BU, bodansky units; SU, sigma units.

[a]Prasad et al.
[b]Sigma Technical Bulletin No. 104, August 1961. (Normal values: adults, 0.8 to 2.3 SU/ml; children, 2.8 to 6.7 SU/ml.)

From Prasad, A. S., in *Zinc Metabolism*, Prasad, A. S., Ed., Charles C Thomas, Springfield, Ill., 1966, 269. With permission.

TABLE 4

Plasma, Erythrocyte, and Urinary Zinc in Normals vs. Dwarfs

	Plasma Zn[a] (μg %)	RBC Zn[a] (μg/ml)	Urinary Zn[a] (μg/day)	Urinary creatinine[a] (g/day)	Urinary creatinine[a] (mg/day/kg)
Normals (9)[b]	108 ± 8	13.4 ± 0.8	613 ± 93	1.42 ± 0.3	22.0 ± 5
Dwarfs (8)	73 ± 6	10.5 ± 0.95	395 ± 46	0.63 ± 0.17	21.9 ± 4

Note: ±, standard deviation.

[a]The difference between normals and dwarfs is statistically significant ($p < 0.01$).
[b]The number in parentheses indicates the number of subjects included in each group.

From Prasad, A. S., in *Zinc Metabolism,* Prasad, A. S., Ed., Charles C Thomas, Springfield, Ill., 1966, 270. With permission.

TABLE 5

Trace Elements in Serum

	Normals	Dwarfs
Copper (μg %)[a]	125 ± 22 (19)[b]	142 ± 24 (17)
Magnesium (meg/l)	1.92 ± 0.19 (17)	1.82 ± 0.12 (12)
Iron (μg %)[c]	87 ± 14 (19)	35 ± 14 (17)
UIBC (μg %)[c]	250–400 (19)	360–650 (17)

Note: UIBC, unsaturated iron-binding capacity; ±, standard deviation.

[a]The difference between normals and dwarfs is statistically significant ($p < 0.05$).
[b]Numbers in parentheses indicate the number of subjects included for each determination.
[c]The differences between normals and dwarfs are statistically significant ($p < 0.01$).

From Prasad, A. S., in *Zinc Metabolism,* Prasad, A. S., Ed., Charles C Thomas, Springfield, Ill., 1966, 270. With permission.

people of Kharga are similar to those in the delta region. Therefore, a field study was conducted in this oasis.[41] Sixteen patients with hypogonadism and dwarfism but only mild anemia were studied. None of them had schistosomiasis or hookworm infections and serum concentrations of iron and zinc were low in the group.

Patients in the delta region of Egypt had both hookworm infection and schistosomiasis, which caused blood loss. Since red blood cells are rich in both iron and zinc, these parasitic infestations are important contributing factors in the production of iron and zinc deficiencies. In Kharga, parasitic infestations did not seem to be responsible for these deficiencies. An analysis of a water sample from an artesian spring, which was the principal source of water for the Kharga villagers, revealed an iron concentration of 317 μg % and a zinc content of 1.8 μg %. In Cairo, the iron and zinc concentrations of drinking water were 7 and 40 μg %, respectively. Although the food consumed by the subjects in both the delta and the oasis villages was similar, the latter probably derived a significant amount of iron, but not zinc, from their water source.

In Egypt and China, dwarfism and hypogonadism have been attributed to liver dysfunction due to schistosomiasis. However, the existence of such patients in Kharga and Iran, where schistosomiasis was absent and liver cirrhosis was not present, indicated that these factors per se were not responsible for these clinical findings. The study also indicated that severe anemia and iron deficiency were not necessary factors for growth retardation and hypogonadism. Furthermore, as discussed earlier, iron deficiency in animal and human subjects does not cause growth retardation and hypogonadism.

As pointed out earlier, no evidence was obtained that indicated a deficiency of other essential trace elements or vitamins. In view of the above findings and the similarity between the clinical features of dwarfism and hypogonadism and those seen in several species with zinc deficiency, it was a reasonable hypothesis to attribute

TABLE 6

Various Elements in Hair (ppm)[a]

No. dwarfs	Calcium	Copper	Nickel	Manganese	Chromium	Cobalt	Molybdenum	Vanadium
1	1000	25	0.3	1	0.3	0.2	0.2	0.10
2	385	22	8.0	3	0.5	0.2	0.2	0.20
3	1550	24	1.0	1	0.5	0.2	0.3	0.10
4	400	22	1.0	5	0.4	0.2	0.2	0.10
5	440	22	0.6	1	0.4	0.2	0.2	0.05
Egyptian controls (12, range)	410–1050	22–32	1–6	1–7	0.2–0.3	0.2	0.2	0.08–0.45
American[b] (90, range)	60–400	14–40	0.3–2	0.2–0.7	0.3–0.8	0.2	0.2	0.02–0.13

[a]All hair samples were collected at the same time. Determinations were done by National Spectrographic Laboratories, Inc., Cleveland, Ohio.
[b]Strain, W. H., personal communication.

From Prasad, A. S., in *Zinc Metabolism,* Prasad, A. S., Ed., Charles C Thomas, Springfield, Ill., 1966, 271. With permission.

the dwarfism and hypogonadism in these subjects to a deficiency of zinc.

It must be emphasized that the anemia in all cases was hypochromic, microcytic in type; related to iron deficiency; and was completely corrected by oral administration of iron salts.

The constant finding of hepatosplenomegaly deserves brief comment. As indicated earlier, we were unable to account for this on the basis of liver disease. This leaves three possibilities: anemia, zinc deficiency, or a combination of these. From our studies, no conclusion could be reached as to its pathogenesis at this time. However, in each case the size of the liver and spleen decreased markedly following treatment with zinc.

The studies in Egypt included only male subjects. Female patients refused to be examined. Furthermore, our facilities at U.S. Naval Medical Research Unit No. 3 (NAMRU-3), Cairo, Egypt, did not allow us to study female subjects. Thus, one could not be certain if this syndrome affected females. However, it is possible that males are more susceptible to zinc deficiency, inasmuch as zinc seems to be concentrated in male genital tracts whereas it is not preferentially accumulated in female genital tracts. Thus, one could have predicted that zinc deficiency in females manifesting growth retardation was probably prevalent, as was later noted in Iran.[46]

The endocrine abnormalities in these patients resembled those of idiopathic hypopituitarism.[35] Growth failure and hypogonadism were the most outstanding features. In addition, some cases showed a decreased pituitary adrenocorticotrophic hormone reserve and an abnormal oral glucose tolerance. Hypothyroidism was not present. Some of these abnormalities have also been described in zinc-deficient animals; therefore, they are probably characteristic of the organism rendered zinc-deficient at a period of expected rapid growth.

Further studies in Egypt showed that the rate of growth was greater in patients who received supplemental zinc as compared to those who received iron instead or those receiving only an animal protein diet consisting of bread, beans, lamb, chicken, eggs, and vegetables.[35] Pubic hair appeared in all cases within 7 to 12 weeks after zinc supplementation was initiated. Genitalia size became normal, and secondary sexual characteristics developed within 12 to 24 weeks in all patients receiving zinc. On the other hand, no such changes were observed in a comparable length of time in the iron-supplemented group or in the group on an animal protein diet. Thus, the growth retardation and gonadal hypofunction in these subjects were related to zinc deficiency. The anemia was due to iron deficiency and responded to oral iron treatment.

In 1966, Coble, Schulert, and Farid[42] reported a follow-up study of patients with dwarfism and hypogonadism from a Kharga oasis that was originally studied by Prasad et al.[41] in 1963. Three years later a majority of those dwarfs

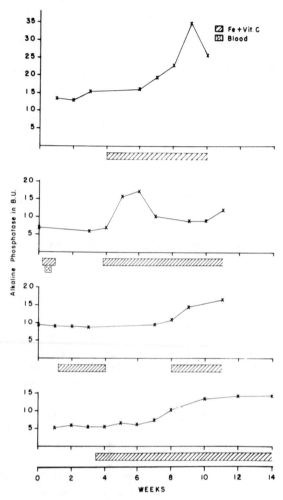

FIGURE 5. Changes in alkaline phosphatase in four dwarfs from delta villages of Egypt, following hospitalization. (From Prasad, A. S., Miale, A., Farid, Z., Schulert, A., and Sandstead, H. H., *J. Lab. Clin. Med.*, 61, 531, 1963. With permission.)

showed an increase in growth and gonadal development without receiving any specific treatment. The plasma zinc concentrations in these patients in 1965 were not altered as compared to their levels in 1962. The authors concluded that these cases merely represent examples of delayed maturation. In another paper, Coble et al.[42a] demonstrated low plasma zinc levels in normal rural male Egyptians. These results were interpreted to show a lack of relationship between growth and status of zinc in the human body; therefore, the essentiality of zinc in human nutrition was questioned. It is clear from these reports[42,42a] that the dwarfs from the Kharga oasis who show increased growth (3 years later) and hypozincemia clearly demon-

strated delayed maturation. Up to the ages of 18 to 19 they failed to attain heights that one would expect to see in Egyptians of an upper socioeconomic level and in normal Americans who showed normal plasma zinc levels. The same was true for "normal" rural Egyptian males, who showed delayed maturation and smaller ultimate statures as compared to their upper socioeconomic counterparts in Cairo.[42,42a] Racially and culturally speaking, the normal subjects from Cairo, belonging to an upper socioeconomic group, are the same as rural "normals." Thus, the so-called rural normals had abnormal growth patterns. Therefore, these observations could be interpreted to show that a low plasma zinc level indeed correlated with slower growth rate and delayed maturation as reported by Coble, Schulert, and Farid[42] and Coble et al.[42,42a] Various degrees of zinc deficient states may exist in given populations. The cases studied by Prasad et al.[41] were selected from the villages as examples of severe zinc deficiency. A milder form may manifest itself only by delayed maturation and slower growth rates.

Carter et al.[43] administered zinc, iron, or placebo as supplements for 5½ months to different groups of adolescent village school boys between the ages of 11 and 18. No differences between the heights, weights, or sexual developments were observed among the three groups. Their studies were not well designed. The subjects for such a study must be paired according to age, height, and sexual development inasmuch as the growth rates of children and adolescents normally differ greatly from one age group to another. This was an important factor responsible for their negative results. One other possibility should be considered in accounting for the negative results in the Cairo study.[43] The supplementation program only lasted for a total period of 5½ months, which under the dietary conditions of the villagers may not have been adequate to see any difference in growth and gonads.

A similar study was carried out in Iran.[44] The subjects were matched according to age, height, and sexual development prior to supplementation with placebo, iron, or zinc. These boys were between 12 to 14 years of age and received supplements for two 5-month periods, with a 7-month interval during which they received no supplements. In this study, the zinc-supplemented group showed significant sexual development and a significant increase in the cortical thickness of the metacarpal bone, thus providing evidence for the important role of zinc in growth and gonadal development.

A more recent study from Iran included 50 13-year-old school boys for supplementation studies.[45] A high protein-vitamin liquid supplement was used to provide all essential micronutrients. Group A received placebo, while groups B and C received a protein-vitamin liquid supplement. Group C received 40 mg of zinc, whereas Group B received none. This supplementation program was carried out for two 9-month periods (no supplementation was given during the summer months since the schools were closed). The zinc-supplemented boys showed greater gains in height and weight than the boys who received identical supplementation minus zinc. Also, the zinc-supplemented group showed a higher proportion of developed genitalia at the end of the study.

These studies clearly demonstrate that zinc is a principal limiting factor in the nutrition of children when the intake of unleavened whole-meal bread is high. It is clear that the 25 to 28 mg of zinc supplement used in the previous studies was not adequate and that 40 mg of zinc, as used in the last study,[45] may have also been marginal since plasma zinc levels were still low. In contrast, when our subjects in Egypt received only 18 mg of supplemental zinc with a diet containing adequate animal protein and calories, response to zinc was rapid and obvious. Thus, requirements of zinc under different dietary conditions vary widely, and this must be considered in order to correct zinc deficiency in a given population.

Halsted et al.[46] recently published the results of their study involving a group of 15 men who were rejected at the Iranian Army Induction Center because of "malnutrition." Two women, 19 and 20 years old, were also included. A unique feature was that all were 19 or 20 years old. Their clinical features were similar to those of zinc-deficient dwarfs reported earlier by Prasad et al.[19] and Sanstead et al.[23] They were studied for 6 to 12 months. One group was given a well-balanced, nutritious diet containing ample animal protein plus a placebo capsule. A second group was given the same diet plus a capsule of zinc sulfate containing 27 mg of zinc. A third group was given the diet without additional medication for 6 months, followed by the diet plus zinc for another 6-month period. The two women lived in the

house of one of the investigators (HAR) and received the same treatment and observation program.

The development in subjects receiving the diet alone was slow while the effect of height increment and onset of sexual function was strikingly enhanced in those receiving zinc. The zinc-supplemented boys gained considerably more height than those receiving the ample protein diet alone. The zinc-supplemented subjects showed evidence of early onset of sexual function, as defined by nocturnal emission in males and menarchy in females. The two women described in this report represented the first cases of dwarfism in females due to zinc deficiency.[46]

The prevalence of zinc deficiency in human populations throughout the world should briefly be mentioned. Clinical pictures similar to these reported in zinc-deficient dwarfs have been observed in many countries such as Turkey, Portugal, and Morocco.[47] Zinc deficiency should also be prevalent in other countries where primarily cereal proteins are consumed by the population. Clinically, perhaps it is not very difficult to recognize extreme examples of zinc-deficient dwarfs in a given population, but it is the marginally deficient subjects who present great difficulties and only future studies can provide insight into this problem. It is now becoming clear that nutritional, as well as conditioned deficiency of zinc may complicate many disease states.

Research on the nutritional status of zinc in infants and young children has been very limited. However, the importance of adequate zinc nutrition in the young is apparent from data on other mammals. Dietary zinc requirements for the young are relatively high as compared with those of mature animals of the same species, and the effects of dietary insufficiency in infants and young children are particularly severe. Furthermore, several of the major features of zinc deficiency (such as growth retardation and impaired learning ability) are peculiar to the young animal.

Zinc Deficiency in Preadolescent School Children in Iran

A clinical syndrome similar to that of "adolescent nutritional dwarfism" has been identified in younger children in Iran, though failure of sexual maturation is not evident prior to adolescence. Clinical features include anemia, hepatosplenomegaly, and growth retardation of undetermined etiology. This syndrome is most common in small, rural communities in which there is also a high incidence of adolescent nutritional dwarfism. Eminians et al.[48] found that mean plasma zinc levels of children with this syndrome were significantly lower than those of normal children in the same village and normal suburban children of the same age (Table 7).

The similarities of this syndrome to that of adolescent nutritional dwarfism and the presence of low serum zinc levels suggest that zinc deficiency contributed to the poor growth of these children. Their diets, which contain large quantities of phytate and fiber, are known to be deficient in available zinc. Unfortunately, no studies of dietary zinc supplementation have been reported for these preadolescent children, and the incidence of this syndrome in younger age groups is unknown.

Zinc Deficiency Associated with Generalized Malnutrition

Plasma zinc levels have been measured in infants and young children suffering from kwashiorkor in Cairo,[49] Pretoria,[50] Cape Town,[51] and Hyderabad.[52] At the time of hospital admission, levels were very low for all four locations (Table 8), but the hypozincemia could be attributed at least in part to hypoalbuminemia. During the subsequent 8 weeks, plasma zinc levels increased, and in Cape Town and Hyderabad they reached normal control levels. However, in Cairo and Pretoria, plasma zinc levels remained significantly below normal at a time of "clinical cure," when total serum protein and serum albumin levels were normal. This persistent hypozincemia suggests that zinc deficiency is associated with kwashiorkor in some geographical locations. The clinical significance of this deficiency has not been defined but may, for example, have contributed to the growth failure[49] and the incidence of skin ulceration.[51] Following recovery from kwashiorkor, children in some areas of the world (including Egypt) are likely to receive a diet, with inadequate zinc; thus, a deficiency of this nutrient may persist indefinitely.

In Cape Town the zinc concentration in the liver, but not in the brain, heart, or muscle, of children dying from kwashiorkor was significantly lower than normal. Plasma zinc levels were also low in marasmic infants.

Secondary Zinc Deficiency in Children

Though data are limited, some children are subjected to increased risk from zinc deficiency secondary to excessive excretion of this element or to intestinal malabsorption. Hypozincemia has been reported in association with celiac disease,[53] disaccharidase deficiency,[54] and cystic fibrosis.[55] The incidence of low hair zinc levels (more than 2 SD below the normal adult mean) in children with cystic fibrosis is ten times that of normal children. Symptomatic zinc deficiency has been detected in patients with regional enteritis.[56]

An increased risk of zinc deficiency is associated with the therapeutic use of synthetic oral[57] and intravenous[58] diets unless these preparations are zinc supplemented. A wide range of zinc concentrations has been found in parenteral hyperalimentation solutions, even for the same commercial preparation. However, the majority of analyzed[73] solutions have had zinc concentrations less than 10 $\mu g/100$ ml, which is inadequate for the infant and young child maintained on total parenteral hyperalimentation. An additional 20 to 40 μg of zinc intravenously per kilogram body weight per day has been found necessary to reverse the hypozincemia resulting from the use of solutions not supplemented with zinc.[59]

TABLE 7

Serum Zinc Levels of Preadolescent Iranian Children

Group cases[a]	Serum Zn, $\mu g/ml$ (mean ± SE)	Hair Zn, $\mu g/g$ (mean ± SE)
Village males symptomatic (17)	0.72 ± 0.06	141 ± 14
Village males control (12)	0.85 ± 0.06	163 ± 22
Urban males control (11)	0.94 ± 0.12	
Village females symptomatic (7)	0.78 ± 0.07	136 ± 14
Village females control (16)	0.95 ± 0.06	172 ± 12
Urban females control (10)	0.93 ± 0.06	222 ± 41

[a]Numbers in parentheses indicate the number of subjects included for each determination.

From Eminians, J., Reinhold, J. G., Kfoury, G. A., Amirhakimi, G. H., Sharif, H., and Ziai, M., *Am. J. Clin. Nutr.*, 20, 734, 1967. With permission.

TABLE 8

Plasma Zinc Levels in Kwashiorkor

Location	Time of admission[a]	"Clinical cure" (3–8 weeks)	Controls
Cairo[49]	42 ± 14.5[b] (37)	68 ± 15.5 (8)	89 ± 6.0
Pretoria[50]	62 ± 20.9 (29)	76 ± 16.5 (9)	113 ± 14.7
Cape Town[51]	42 ± 15 (15)	86 ± 14 (15)	90 ± 28
Hyderabad[52]	41 ± 8.1 (28)	91 ± 9.8 (14)	102 ± 12.4

[a]Numbers in parentheses indicate the number of subjects included for each determination.
[b]μg Zn per 100 ml ± SD.

From Hambidge, K. M. and Walravens, P. A., in *Trace Elements in Human Health and Disease,* Vol. 1, Prasad, A. S., Ed., Academic Press, New York, 1976, 23. With permission.

TABLE 9

Growth, Appetite, and Taste Acuity of Children Aged 4 to 16 Years with Hair Zinc Levels <70 ppm

Age (years)	Sex	Hair zinc, (ppm)	Percentiles Weight	Height	Poor appetite	Hypogeusia
5	F	10	<3rd	<10th	–	+
7	M	18	3rd	50th	+	+
4	F	27	10th	3rd	+	
7	F	43	<3rd	<3rd	+	+
5	F	56	10th	<10th	+	+
7	M	58	75th	75th	–	
13	M	62	<10th	3rd	–	
6	M	63	<3rd	<3rd	+	+
5	M	64	<3rd	<3rd	+	
5	M	67	25th	10th	+	–

From Hambidge, K. M. and Walravens, P. A., in *Trace Elements in Human Health and Disease,* Vol. 1, Prasad, A. S., Ed., Academic Press, New York, 1976, 25. With permission.

Nutritional Zinc Deficiency in Children in the United States

In 1972, a number of Denver children were reported to exhibit evidence of symptomatic zinc deficiency.[60] These children were identified as a result of a survey of trace element concentrations in the hair of apparently normal children from middle and upper income families. Out of 132 children 4 to 16 years of age, 10 had hair zinc concentrations less than 70 ppm (or more than 3 SD below the normal adult mean). This was an unexpected finding in normal children who had had no recent or chronic illness, and consequently additional data were obtained.

Eight of these ten children were found to have heights at or below the tenth percentile on the Iowa Growth Charts (Table 9), although they were not preselected according to height. There was no apparent cause for the relatively poor growth, which could not be explained on a familial basis. Growth retardation is one of the earliest manifestations of zinc deficiency in the young animal, and the correlation between the low hair zinc levels and low growth percentiles in these children suggested a causal relationship.

Anorexia is another prominent, early feature of zinc deficiency in animals, and most of the children with low hair zinc levels in the original Denver study also had a history of poor appetite. In particular, the consumption of meats was very limited despite access to larger quantities. As animal products provide the best source of zinc, it is quite possible that the dietary zinc intake of these children was inadequate.

Another typical feature manifested by these children was hypogeusia (impaired taste acuity). Tests were repeated 1 to 3 months after commencing dietary zinc supplementation (0.2 to 0.4 mg Zn per kilogram body weight/day), and the results showed taste acuity to be normal in each case. This improvement could not be attributed to a placebo effect.[61] Hair zinc levels increased concurrently with the improvement in taste acuity. The rapid response to very small quantities of supplemented zinc provided strong evidence for a preexisting nutritional deficiency of this metal. Therefore, it was concluded that these otherwise normal children were suffering from a dietary deficiency of zinc sufficient to impair taste acuity and possibly also to disturb appetite and limit the rate of growth.

Although it was long considered impossible, suboptimal zinc nutrition has recently been calculated to pose a risk for substantial sections of the population of the U.S.[62] Those in particular danger include subjects whose zinc requirements are relatively high (for example, people subsisting on low income diets and those experiencing rapid growth). The original Denver study did not include children from low income families. Subsequently, studies have been undertaken on 29 young children enrolled in the Denver Head Start program whose heights, with only three exceptions, were below the third percentile. The mean hair zinc level of this group was significantly lower than that of children of the same age from middle and upper income families (43% had hair zinc levels less than 70 ppm). In addition, the mean plasma zinc concentration and the mean rate of zinc secretion in parotid saliva were significantly lower than normal. Satisfactory measurements of taste acuity were achieved in six children, and objective hypogeusia was present in each case.

It is interesting to note that 33% of these children had low levels of serum vitamin A (<20 μg %). The investigations were undertaken 3 months after enrollment in the Head Start program (during that interval Head Start meals had been designed to supply adequate quantities of carotene). Although zinc depletion in the rat prevents normal release of vitamin A from the liver (resulting in low serum levels of this vitamin),[63] it is not known if this is a cause of low serum vitamin A levels in children. These children had received supplemental iron for 3 months prior to this study and were not anemic. Total serum protein and serum albumin levels were normal in each case.

Zinc Nutrition and Deficiency in Infants
Low Levels of Zinc in the Plasma and Hair of U.S. Infants

Hair and plasma zinc levels are exceptionally low in infants in the U.S. as compared to other age groups including the neonate, older children, and adults.[60,64] In the Denver survey of hair trace element concentrations (discussed above), there was a mean hair zinc level of only 74 ± 8 ppm (mean ± SE) for the 26 apparently normal infants, aged 3 to 12 months, as compared with 174 ± 8 ppm for neonates and 180 ± 4 ppm for adults. Of these infants from middle and upper income families, 54% had hair zinc levels less than 70 ppm. Exceptionally low hair zinc levels have also been found in infants residing in Dayton, Ohio.[65,66] It appears unlikely that these low levels can be regarded as entirely normal for this age, since levels are not equally low in other countries where comparable data have been obtained. For example, in Thailand the mean hair zinc level of 15 infants from upper economic level homes was 202 ± 26 ppm; no source of external contamination could be identified in Bangkok to account for these higher levels. Hair zinc concentrations of adults in Thailand were closely comparable to those of Denver adults;[67] however, age-related differences in the younger children as well as infants were very different from those in Denver (Figure 6). These discrepancies indicate that the variations are not directly related to age and that the low levels in

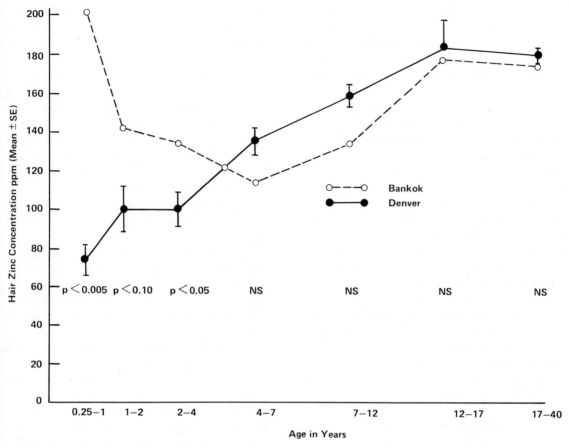

FIGURE 6. Mean hair zinc concentrations of normal subjects from middle and upper income families in Denver and Bangkok. (From Hambidge, K. M. and Walravens, P. A., in *Trace Elements in Human Health and Disease,* Vol. 1, Prasad, A. S., Ed., Academic Press, New York, 1976, 28. With permission.)

Denver infants cannot necessarily be accepted as physiological. Similar considerations apply equally to the low plasma zinc levels reported for infants in this country.[64] In Sweden[68] and Germany,[53] plasma zinc concentrations in infants have been reported to be no lower than those for adults.

Several factors, including difficulty in achieving positive zinc balance in early postnatal life[69,70] and a "dilutional" effect of rapid growth, may contribute to zinc depletion in infants. A unique factor in the U.S. that may contribute to zinc deficiency is the low concentration of this metal in certain popular infant milk formulae. Formulae in which the protein content of the original cow's milk is reduced to levels comparable to those of human milk have a parallel reduction in zinc content. Unless subsequently supplemented with zinc, these formulae typically have a zinc concentration of less than 2 mg/l or approximately half that of cow's milk and human milk, during

the first 2 months of lactation. Zinc supplementation of all of these formulae has not been routine, but it is likely to become a universal practice in the near future, following the recent recommendations of the Food and Nutrition Board of the National Academy of Sciences.[70a] Indirect support for the possible role of these formulae in the low hair zinc levels of U.S. infants has been derived from measurements of hair zinc concentrations in English formula-fed infants. The mean hair zinc level of these infants was 124 ± 12 ppm, which, though lower than adults, was 68% higher than the Denver mean ($p < 0.005$).

Symptomatic Zinc Deficiency in Infants

The majority of infants with low plasma and hair zinc levels have not shown any detectable signs of zinc deficiency. However, it appears that those at the lower end of the spectrum of zinc depletion (as manifested by low levels of hair and plasma zinc) and perhaps those who remain moderately depleted for a prolonged period of time, do develop symptomatic zinc deficiency. In the original Denver study, 8 out of 93 infants and children less than 4 years old had hair zinc levels less than 30 ppm, and 6 of these manifested declining growth percentiles and poor appetite.[60] It was also noted that the high percentiles of the older children with hair zinc levels less than 70 ppm first declined during infancy; thus, if their poor growth resulted from an insufficiency of this nutrient, it must have commenced during infancy. It is conceivable that once a deficiency state has been established, the resulting anorexia may tend to perpetuate the deficiency state. Experience with a number of infants who had low levels of zinc in plasma and hair and responded favorably to zinc supplementation indicates that some cases of failure to thrive in infancy may be caused by zinc deficiency. Anorexia has been a prominent feature in these infants, and one case manifested a bizarre form of pica which improved dramatically following zinc supplementaion.[71] However, currently there is a lack of controlled studies of dietary zinc supplementation in these infants.

DIAGNOSIS OF ZINC DEFICIENCY IN MAN

The laboratory criteria for the diagnosis of zinc deficiency are not completely well established.

The response to therapy with zinc is probably the most reliable index for diagnosing a zinc-deficient state in man.

Low levels of plasma zinc have been observed in patients with many disease states (Table 10). At present, it is not certain whether a low level of plasma zinc is indicative of "zinc deficiency" in all of these conditions. It seems likely that low plasma zinc reflects an impaired zinc nutriture in many, while in others it may reflect a shift of zinc from the plasma to another body pool. In acute infections and myocardial infarction, one may see a drop in the plasma zinc level that is probably related to a shift of zinc from the plasma pool to the tissues.

The concentration of zinc in hair appears to reflect chronic zinc nutriture. Thus, if the hair has been growing at a reasonable rate, it is a useful index of chronic zinc status in the body. However, hair zinc does not reflect changes in the status of zinc on an acute basis. Similar remarks apply with respect to zinc in the red blood cells. Inasmuch as these cells are not turning over rapidly, their zinc content cannot be expected to reflect acute changes taking place in the body as a result of altered zinc nutriture. In terms of the assessment of zinc status in man, perhaps leukocyte or platelet zinc determination may prove to be useful; however, presently only limited data are available and, as such, future investigations are needed.

Excretion of zinc in urine decreases as zinc

TABLE 10

Causes of Zinc Deficiency in Human Subjects

Dietary	Excessive intake of phytate, fiber, polyphosphates, clay and laundry starch, and alcohol
Malabsorption	Pancreatic insufficiency, steatorrhea, gastrectomy, intestinal mucosal disease
Cirrhosis of the liver	
Renal disease	Nephrotic syndrome, renal tubular disease
Renal dialysis	
Chronically debilitating diseases	Neoplastic diseases, chronic infections
Psoriasis (skin loss)	
Burns	
Parasitic infections (chronic blood loss)	
Iatrogenic	Penicillamine therapy, total parenteral nutrition, surgical trauma
Genetic	Sickle cell disease, acrodermatitis enteropathica
Pregnancy	

deficiency progresses.[72] However, this test requires a complete collection of urine on a 24-hr basis. Although most cases of zinc deficiency in man are associated with hypozincuria, there may be certain exceptions. Cirrhosis of the liver and sickle cell disease are usually associated with hyperzincuria, although deficiency of zinc may be present. The mechanism of hyperzincuria in these diseases remains to be elucidated. Hyperzincuria has also been observed to occur in certain renal diseases and infections as well as following injury, burns, and acute starvation.[75,76]

The metabolic balance study, turnover rate, and 24-hour exchangeable pool for zinc (as measured by zinc-65) may provide additional tools for assessing zinc status in man.[23,77,78] The zinc balance study, although difficult, may provide a good basis for assessment of zinc status, and a positive retention of zinc by human subjects would be indicative of zinc deficiency. By using zinc-65, it was shown in zinc deficiency that plasma zinc turnover was increased, 24-hour exchangeable pool was decreased, and cumulative excretion of zinc-65 in urine and stool was low.[23] Unfortunately, the half-life of zinc-65 is 245 days and, as such, this isotope cannot be recommended for routine clinical use.

Urinary sulfate excretion following an injection of cystine-^{35}S to zinc-deficient rats is enhanced presumably due to its inability to utilize sulfur-containing amino acids for protein synthesis.[79,80] Whether or not a similar approach will be fruitful in man remains to be tested. Further work is undoubtedly needed in order to establish tests that may be considered definitive for diagnosing of zinc deficiency in man.

The activities of certain zinc-dependent enzymes may be measured in the plasma. A good correlation between zinc status and the activity of ribonuclease (RNase) has been observed.[120] An increase in the activity of plasma alkaline phosphatase following zinc supplementation to subjects with zinc deficiency has also been observed.[18,35, 72,81] This may provide additional evidence retrospectively, for diagnosing zinc deficiency in man.

CONDITIONED DEFICIENCY OF ZINC IN MAN

Nutritional Factors

All of the conditioned dietary factors listed in Table 10, with the exception of alcohol, probably influence zinc nutriture by making zinc unavailable for absorption. The formation of insoluble complexes with calcium and phytate in the alkaline intestinal environment has been shown to markedly decrease the availability of zinc for intestinal absorption in experimental animals.[82] Dietary fiber has also been shown to decrease the availability of zinc for intestinal absorption by man.[83] It seems probable that both dietary phytate and fiber may contribute to the occurrence of zinc deficiency in populations that subsist largely on bread and other foods rich in fiber and phytate.[83,84]

The effects on human zinc nutriture of ethylenediaminetetraacetate (EDTA) and polyphosphates which may be added to foods during processing are unknown. In the rat, EDTA is known to increase zinc absorption.[82] However, it also increases excretion of zinc in urine.[85] The net effect of a low concentration of EDTA on animals fed zinc-deficient diets containing soybean protein is beneficial rather than harmful.[86]

Geophagia is known to be associated with the occurrence of dwarfism in Iran.[46] Clay may complex zinc in much the same manner that it binds iron.[20] However, this has not been carefully worked out. It is unknown whether clay eating by pregnant black women in certain areas of the U.S. has a significant effect on their zinc nutriture. Laundry starch is included in Table 10 because iron deficiency is known to occur in women who consume it in large amounts.

Alcohol induces an increase in the urinary excretion of zinc;[87,88] The mechanism is unknown. A direct effect of alcohol on the renal tubular epithelium may be responsible for hyperzincuria. Acute ingestion of alcohol did not induce zincuria[88,98] in some experiments; however, one group of investigators reported increased urinary zinc excretion following alcohol intake.[90] This effect was evident when complete urine collection was analyzed for zinc during first 3-hr and second 3-hr periods, following ingestion of 6 oz of chilled vodka.[90]

Liver Disease

Vallee et al.[34,91] initially described the abnormal zinc metabolism that occurs in patients with alcoholic cirrhosis. These investigators demonstrated that patients with cirrhosis had low serum zinc; diminished hepatic zinc; and, paradoxically, hyperzincuria. These observations led them to suggest that zinc deficiency in the alcoholic cirrhotic patient may be a conditioned deficiency which was somehow related to alcohol ingestion.

The initial observation that cirrhotic patients had low serum and hepatic zinc levels and hyperzincuria has been corroborated in many laboratories.[88,92-102] Thus, the initial observations seem to be firmly established, although there is some variance of opinion as to how often phenomena occur in cirrhosis and other liver disease, as well as in other clinical entities. Kahn et al.[98] reported that 4 out of 11 patients with viral hepatitis had elevated urinary zinc (normal mean + 2 SD) and 8 out of 10 of these patients had at least one serum zinc that exceeded their normal values by 2 SD. Halsted et al.[100] found that the plasma zinc was 1 to 2 SD below their laboratory's mean normal plasma zinc values in six patients with viral hepatitis. Moreover, it was equally low in other types of active liver disease in addition to alcoholic cirrhosis. Henkin and Smith[103] studied 19 patients with viral hepatitis and found a significantly decreased serum zinc and a normal serum copper level, during the early stages of the illness. As the hepatitis subsided, the serum zinc rose to normal levels and the serum copper rose to significantly elevated levels. Initially, patients had hyperzincuria and hypercupruria. Both urinary abnormalities reverted to normal as the hepatitis subsided. Changes in serum zinc correlated with changes in retinol-binding protein and total serum protein. (Retinol-binding protein is known to transport vitamin A from its storage site in the liver to peripheral tissues.) Vitamin A was also decreased in the blood of these patients.

Liberation of a leukocyte endogenous mediator, which lowers serum zinc and enhances zinc uptake by the liver, may be an additional factor responsible for a decrease in serum zinc level of cirrhotic subjects.[104] To date, no study has taken this factor into consideration.

Normal subjects excreted 322 ± 167 μg (mean $\pm$ 1 SD) urinary zinc per 24 hr, and the serum values were 94 ± 11 μg/100 ml.[88,95] In 53 of 124 alcoholic patients, the mean normal urinary excretion exceeded 700 μg/24 hr. In 21 of 30 alcoholic individuals, serum zinc was less than 70 μg/100 ml.[95] In the majority of the alcoholic patients with hyperzincuria, the abnormal zinc excretion

reverted to normal within 1 to 2 weeks following alcohol abstention and institution of an adequate diet.

The incidence of low serum zinc and hyperzincuria in cirrhosis varies somewhat from one series to the other. Kahn et al.[98] found these abnormalities of zinc metabolism only in severely ill cirrhotic patients who were classified as decompensated.

Patek and Haig[105] showed that some cirrhotic patients had night blindness which did not improve with vitamin A therapy. One may speculate that retinol-binding protein may be zinc dependent, therefore, zinc administration may be required for correction of night blindness in cirrhotic subjects.

Several interesting studies have been directed toward elucidating the role of zinc in the pathogenesis of liver disease. Kahn and Ozeran[106] chronically injected rats with carbon tetrachloride to produce cirrhosis and found that these cirrhotic animals had low serum zinc and diminished levels of hepatic zinc. In another study, Kahn et al.[107] demonstrated enhanced uptake of zinc-65 by the cirrhotic liver of carbon tetrachloride-treated rats with low serum zinc. These studies seemed to indicate a decreased pool size of zinc and thus implied true zinc deficiency. Kinetic studies of zinc-65 in ten human cirrhotics also suggested a diminished pool size and a slower turnover of zinc.[97] Voigt and Saldeen,[108] Saldeen and Brunk,[109] and Saldeen[110] have shown that parenteral zinc salts protect rat liver from damage by carbon tetrachloride. Similarly, Srinivasan and Balwani[111] showed a protective effect of zinc sulfate on the hepatotoxicity of carbon tetrachloride.

These studies suggest that zinc exerts a protective effect on the liver. Chvapil, Ryan, and Zukowski[112] have shown that zinc inhibits lipid peroxidation of membranes and thereby stabilizes them. Such an effect protects hepatic cells and their organelles against injury. Further studies are required to show whether cellular injury results in the zinc loss from the body or whether zinc loss from the body (and the mechanism for the loss) occurs first, resulting in enhanced lability of the cells to injury.

Many patients with cirrhosis of the liver are known to exhibit a markedly enhanced sensitivity to drugs. Hepatic coma may be precipitated by administration of methionine to cirrhotic patients with an Eck fistula.[113] Similarly, elevated blood ammonia seems to be intimately related to the development of hepatic coma. Anthony, Woosley, and Hsu[114] and Hsu and Woosley[115] have shown that zinc-deficient rats have a defect in the metabolism of sulfur-containing amino acids. In their studies, [35]sulfur-tagged methionine, cystine, or taurine was given to both control and zinc-deficient rats. Methionine can be metabolized to cystine, which in turn can be metabolized to taurine. Their results showed that zinc-deficient animals excreted increased amounts of [35]sulfur and ^{35}S-taurine in the urine.

Zinc deficiency may also affect urea synthesis. Thus, abnormalities related to the metabolism of amino acids and ammonia may act in concert to produce the picture of hepatic coma; Zieve, Doizaki, and Zieve[116] have recently shown that large doses of ethanethiol, methanethiol, and dimethyl sulfide produce coma in rats. Large doses of ammonium salts produce the same results. One may hypothesize that zinc deficiency associated with cirrhosis of the liver may result in two different metabolic abnormalities, which in turn may act synergistically to produce hepatic coma.

Gastrointestinal Disorders

Steatorrhea may be the most common mechanism for zinc deficiency in patients with gastrointestinal disease. In an alkaline environment zinc would be expected to form insoluble complexes with fat and phosphate, analogous to those formed by calcium and magnesium. Thus, fat malabsorption due to any cause should result in an increased loss of zinc in the stool. This appears to have been the case in a patient with severe regional enteritis and steatorrhea (23 g of fat per day) studied by Sandstead, Vo-Khactu, and Solomon[74] (Figure 7). The diet he was fed contained 11.83 mg of zinc per day while his stool contained 11.10 mg/day. Zinc deficiency has been reported in another patient with steatorrhea,[54] and low concentrations of plasma zinc have been observed in patients with disease known to cause steatorrhea.[55] Thus, although metabolic studies have been limited, it seems reasonable to presume that steatorrhea may significantly influence zinc nutrition. The formation of zinc soaps and other insoluble zinc chelates may be potentiated in patients with achlorhydria.

Another potential cause of impaired absorption of zinc may be due to a lack of "zinc-binding

factor." Recent studies in experimental animals[117] indicate that a zinc-binding ligand is secreted into the intestinal lumen by the pancreas. The ligand appears to facilitate the uptake of zinc by intestinal epithelial cells. At present, it is not known whether a lack of the zinc-binding ligand in man may have contributed to the occurrence of low plasma zinc concentration in patients with cystic fibrosis (CF) of the pancreas and growth retardation.[55] Kowerski, Blair-Stanek, and Schacter[118] recently isolated a soluble zinc-binding protein from rat jejunal mucosa that is apparently involved in active transport of zinc across the intestinal mucosa. It has not been demonstrated whether or not such a mechanism for zinc absorption exists in man. If a specific protein in the small intestine is indeed needed for zinc absorption in man, one may speculate that a lack of this protein may cause malabsorption of zinc.

Exudation of large amounts of zinc protein complexes into the intestinal lumen may also contribute to the decrease in plasma zinc concen-

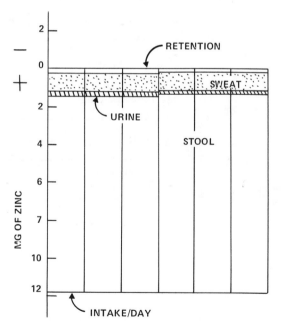

FIGURE 7. Zinc balance of a patient with regional enteritis and steatorrhea. The levels of zinc and urine are based on chemical analysis. The level in sweat is the estimated zinc loss if the patient had one liter of sweat daily. Nearly 90% of the dietary zinc appeared in the stool. (From Sandstead, H. H., Vo-Khactu, K. P., and Solomon, N., in *Trace Elements in Human Health and Disease*, Vol. 1, Prasad, A. S., Ed., Academic Press, New York, 1976, 37. With permission.)

trations which occur in patients with inflammatory disease of the bowel (Figure 8). It seems likely that protein-losing enteropathy due to other causes may also impair zinc homeostasis. Another potential cause of negative zinc balance is a massive loss of intestinal secretions.

Hyperzincuria and Renal Disease

Increased urinary losses of zinc may occur in some patients with cirrhosis and other types of liver disease [34,85,88,94,119] and sickle cell disease.[120] Presumably, the tubular reabsorption of zinc is decreased. One might expect the failure of zinc tubular reabsorption to be mediated by aldosterone. However, a limited study suggests that aldosterone does not influence zinc excretion.[121] Patients with cirrhosis who excrete large amounts of zinc in their urine may have an increased risk of zinc deficiency, particularly if

they consume diets which are limited in protein, since the level of dietary zinc is correlated with dietary protein.[122,123] Steatorrhea in patients with liver disease would be an additional compounding factor in the pathogenesis of zinc deficiency.

The potential causes of conditioned deficiency of zinc in patients with renal disease include proteinuria[124] and failure of tubular reabsorption. In the former instance, the loss of zinc-protein complexes across the glomerulus is the mechanism. In the latter, an impairment in the metabolic machinery of tubular reabsorption due to a genetic abnormality or toxic substances would result in zinc loss. While low plasma zinc concentrations have been described in patients with massive proteinuria,[124] no reports of low plasma levels of zinc in patients with tubular reabsorption defects have appeared in literature.

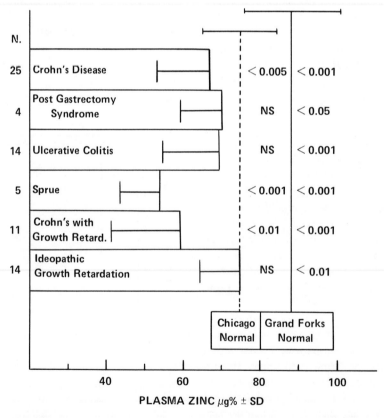

FIGURE 8.   Plasma zinc concentrations of patients with gastrointestinal diseases and idiopathic growth failure. The standard deviations of the means are indicated within each bar and at the top of each normal mean on the right. (From Sanstead, H. H., Vo-Khactu, K. P., and Solomon, N., in *Trace Elements in Human Health and Disease,* Vol. 1, Prasad, A. S., Ed., Academic Press, New York, 1976, 39. With permission.)

In patients with renal failure,[55,125,126] the occurrence of conditioned zinc deficiency may be the result of a mixture of factors, which are presently ill defined. If 1,25-dihydroxycholecalciferol plays a role in the intestinal absorption of zinc, an impairment in its formation by the diseased kidney would be expected to result in malabsorption of zinc. In some individuals with renal failure it seems likely that plasma and soft tissue concentrations of zinc may be "protected" by the dissolution of bone, which occurs as a result of increased parathyroid activity in response to low serum calcium. In experimental animals, calcium deficiency has been shown to cause release of zinc from bone.[127] In some patients who are successfully treated for hyperphosphatemia and hypocalcemia, the plasma zinc concentration may be expected to decline due to the deposition of zinc and calcium in bone.[128,129] Thus, in the latter group in particular a diet low in protein and high in refined cereal products and fat would be expected to contribute to a conditioned deficiency of zinc. Such a diet would be low in zinc.[122,123,130] The patients studied by Mansouri, Halsted, and Gombos,[126] who were treated with a diet containing 20 to 30 g of protein daily and had low plasma concentrations of zinc; they appear to represent such a clinical instance. Presumably the patients of Halsted and Smith[55] (Figure 9) were similarly restricted in dietary protein. In other patients with renal failure whose dietary protein was not restricted, plasma zinc concentrations were not decreased.[125] Patients on dialysis had even higher levels, particularly following dialysis. Apparently, zinc deficiency may not be a problem in patients on dialysis if their dietary consumption of protein is not restricted.

Neoplastic Disease

The occurrence of conditioned zinc deficiency in patients with neoplastic disease will obviously depend on the nature of the neoplasm.[131,132] Anorexia, starvation,[133] and avoidance of foods rich in available zinc are probably important conditioning factors. An increased excretion of zinc subsequent to its mobilization by leukocyte endogenous mediator (LEM)[134] in response to tissue necrosis may be another factor.[135]

Burns and Skin Disorders

The causes of zinc deficiency in patients with burns include losses in exudates.[133,136-138] Starvation of patients with burns is a well-recognized cause of morbidity and mortality. The contribution of conditioned zinc deficiency to the morbidity of burned patients is not defined. Limited studies indicate that epithelialization of burns may be improved by treatment with zinc.[133,137,139] Such a finding is consistent with studies on the beneficial effect of zinc in the treatment of leg ulcers.[140-144] and the well-defined requirement of zinc for collagen synthesis.[145]

In psoriasis, the loss of large numbers of skin cells may result in zinc depletion. The skin contains approximately 20% of the body zinc. Thus, if the loss of epithelial cells is great enough, it is conceivable that the massive formation of new cells could lead to conditioned deficiency. Low levels of plasma zinc have been reported in some patients with extensive psoriasis,[146,147] however, these findings have not been confirmed.[148,149]

Impaired Wound Healing in Chronically Diseased Subjects

In 1966, Pories et al.[150-152] reported that oral administration of $ZnSO_4$ to military personnel with marsupialized pilonidal sinuses was attended by a twofold increase in the rate of reepithelialization. The authors' conclusion that $ZnSO_4$ can promote the healing of cutaneous sores and wounds has been subject of controversy during the past several years.[152a] As summarized in Table 11, clinical investigations by Cohen,[153] Husain,[140] Greaves and Skillen,[154] and Serjeant, Galloway, and Gueri;[141] have substantiated the beneficial effects of $ZnSO_4$ on wound healing, whereas studies by Barcia,[155] Myers and Cherry,[156] Clayton,[157] and Greaves and Ive[158] have failed to demonstrate any therapeutic benefit. Hallböök and Lanner[159] found that the reepithelialization rate of venous leg ulcers was enhanced by $ZnSO_4$ in patients who initially had diminished concentrations of serum zinc; however, they did not find any benefit in patients whose initial measurements of serum zinc were within the normal range. On the other hand, Husain, Fell, and Scott[160] did not observe any relationship between response to zinc treatment of venous leg ulcers and the initial measurement of plasma zinc.

Studies in experimental animals have demonstrated that:

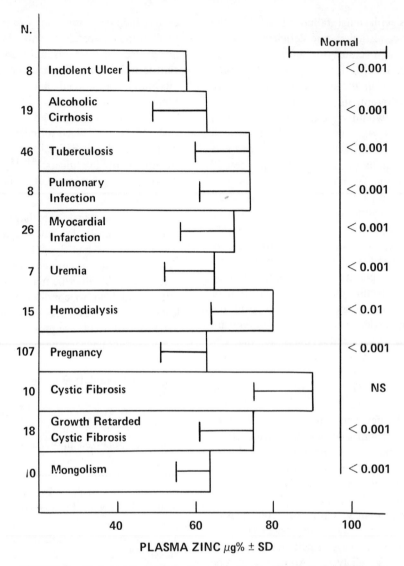

FIGURE 9.   Plasma zinc concentrations of patients with selected diseases. The
number of cases is indicated on the left. The standard deviation from the normal
mean is indicated on the right. The standard deviation from the mean plasma zinc
value for each group of patients is indicated within the bars. (From Halsted, J. A.
and Smith, J. C., Jr., *Lancet*, 1, 322, 1970. With permission.)

1. Healing of incised wounds is impaired in rats with dietary zinc deficiency.[144,161-163]

2. Collagen and noncollagen proteins are reduced in skin and connective tissues from rats with dietary zinc deficiency.[145,164,165]

3. Zinc supplementation does *not* augment wound healing in normal rats.[166,167]

4. Zinc supplementation *does* augment wound healing in chronically ill rats.[168]

These data provided evidence that zinc supplementation may promote wound healing in zinc-deficient patients.

Parasitic Infestations

Parasitic disease which causes blood loss may contribute to conditioned deficiency of zinc. This appears to have been the case in the zinc-responsive "dwarfs" reported from Egypt.[19,35] As red blood cells contain 12 to 14 μg of zinc per milliliter, infections with hookworm and/or schistosomiasis which are severe enough to cause

TABLE 11

Clinical Trials of Oral $ZnSO_4$[a] in Healing of Sores

| Investigator | Date | Lesions | No. of patients | | Observation on reepithelialization rate |
			Controls	Treated	
Pories et al.[150-152]	1966	Pilonidal sinus	10	10	Twofold increase vs. controls
Cohen[153]	1968	Bedsores	0	6	Improvement
Husain[140]	1969	Leg ulcers	52	52[b]	Twofold increase vs. controls
Barcia[155]	1970	Pilonidal sinus	10	10	No influence vs. controls
Greaves and Skillen[154]	1970	Leg ulcers	0	18	Complete healing in 11 patients; partial healing in 7 patients
Serjeant, Galloway, and Gueri[141]	1970	Leg ulcers	17	17[b]	Threefold increase vs. controls
Myers and Cherry[156]	1970	Leg ulcers	16	16[b]	No influence vs. controls
Clayton[157]	1972	Leg ulcers	5	5[b]	Slower in treated vs. controls
Greaves and Ive[158]	1972	Leg ulcers	18	18[b]	No influence vs. controls
Hallböök and Lanner[159]	1972	Leg ulcers	14	13	Fourfold increase vs. controls in patients with low serum Zn; no influence vs. controls in patients with normal serum Zn

[a]$ZnSO_4$, 220 mg three times daily.
[b]Double-blind study.

From Sunderman, F. W., Jr., *Ann. Clin. Lab. Sci.,* 5, 132, 1975. With permission.

iron deficiency will probably contribute to the occurrence of zinc deficiency.

Iatrogenic Cause

Possible iatrogenic causes of conditioned deficiency of zinc include the use of antimetabolites and antianabolic drugs.[169] Treatment with some of these drugs causes patients to feel ill. They become anorectic and may starve. With catabolism of body mass, urinary excretion of zinc is increased.[138] Commonly used intravenous fluids are relatively zinc free. Thus, under normal circumstances, a negative zinc balance should occur in patients who are given antimetabolites, antianabolic agents, or prolonged intravenous therapy.

Failure to include zinc in fluids designed for total parenteral nutrition (TPN) is another example of iatrogenically induced, conditioned deficiency of zinc. A decline of plasma zinc has been observed in some patients given TPN fluids containing less than 1.25 mg of zinc daily.[74]

Zinc deficiency has recently been reported to occur in patients following penicillamine therapy for Wilson's disease.[170] The manifestations con-

sisted of parakeratosis, "dead" hair and alopecia, keratitis, and centrocecal scotoma. Following supplementation with zinc, several clinical manifestations were reversed.

Diabetes

Some patients with diabetes mellitus have been found to have increased urinary losses of zinc,[171,172] the mechanism is unknown. Presumably, some of them may become zinc deficient although in general the plasma zinc is not affected.[172] Perhaps the failure of some diabetics to heal ulcers of the feet (and elsewhere) is related to zinc deficiency. Healing of such ulcers in diabetes has been reported subsequent to zinc therapy.[133]

Collagen Diseases

In patients with inflammations such as rheumatoid arthritis, lupus erythematosus,[173] infection, or injury, two factors may lead to zinc deficiency. The loss of zinc from catabolized tissue[138] and mobilization of zinc by leukocyte endogenous mediator (LEM)[134] to the liver and its subsequent excretion in the urine may account for the conditioned zinc deficiency.

Pregnancy and Oral Contraceptives

Plasma concentration of zinc decreases in human pregnancy.[174] Presumably, the decrease partly reflects the uptake of zinc by the fetus and the other products of conception. It has been estimated that pregnant women must retain approximately 750 μg of zinc per day for growth of the products of conception during the last two thirds of pregnancy.[62] Thus, when zinc deficiency occurs in pregnancy, a conditioning factor is the fetus's demand for zinc. Studies in the rat suggest that the placenta actively provides zinc to the fetus.[139] If the diet of pregnant women does not include liberal amounts of animal protein, the likelihood of conditioned deficiency of zinc is increased because zinc is probably less available through foods derived from grains and other plants.[175] The possible importance of zinc deficiency in human pregnancy is implied by the observations reported by Hurley,[176] Caldwell et al.,[177] and Halas et al.[178] Zinc deficiency in pregnant rats was shown to cause fetal abnormalities, behavioral impairment in the offspring, and difficulty in parturition.[176-178]

Caldwell et al.[177] were the first to show that even a mild zinc deficiency in rats in both prenatal and postnatal nutrition had a profound influence on behavior potential despite an apparently adequate protein level in the diet. Recently, it has been observed that zinc deficiency in fetal and suckling rats results in adverse biochemical effects in the brain.[178] These adverse effects are apparently not readily reversible, as behavioral testing of nutritionally rehabilitated 60- to 80-day-old male rats has shown that they performed poorly on a Tolman-Honzik maze[179] when compared to pair-fed and ad libitum-fed control rats. These findings suggest that zinc deficiency during the critical developmental period of the rat brain induces almost irreversible abnormalities, which are manifested by impaired behavioral development.

Hurley[176] has shown that short-term depletion of zinc in maternal rats results in a wide variety of congenital anomalies in the offspring. In view of the important role of zinc in nucleic acid metabolism, Hurley and Schrader[180] proposed that impaired DNA synthesis in zinc-deprived embryos prolongs the mitotic cycle and reduces the number of normal neural cells, which leads to malformations of the central nervous system. It is tempting to speculate that the exceptionally high rates of congenital malformations of the central nervous system (as reported from the Middle East)[181] may be caused by maternal zinc deficiency.

Human amniotic fluid has been shown to contain an inorganic bacterial inhibitory component which was identified as zinc.[182,183] The average zinc concentration in amniotic fluid was found to be 0.44 μg/ml. The phosphate concentration of amniotic fluid appeared to determine the expression of zinc inhibitory activity. The average phosphate concentration was found to be 92 μg/ml. For the 22 fluids tested, a phosphate-to-zinc ratio of 100 or less predicted a bactericidal fluid. A ratio between 100 to 200 predicted a bacteriostatic fluid, whereas a ratio greater than 200 was noninhibitory. These observations may have clinical significance if maternal zinc nutrition indeed affects the concentration of zinc in the amniotic fluid.

Plasma zinc is known to decrease following use of oral contraceptive agents.[184-186] Our recent data indicate that whereas the plasma zinc may decline, the zinc content of the red blood cells increases as a result of administration of oral contraceptive agents. This phenomenon may merely indicate a redistribution of zinc from the plasma pool to the red cells. Alternatively, oral contraceptive agents may enhance carbonic anhydrase (a zinc metalloenzyme) synthesis, thus increasing the red cell zinc content.

Genetic Disorders
Sickle Cell Disease

Deficiency of zinc in sickle cell disease has recently been recognized.[120] Certain clinical features common in some sickle cell anemia patients are also common in zinc deficient patients. The latter was reported from the Middle East.[72,120] These symptoms include delayed onset of puberty and hypogonadism in the males (characterized by decreased facial, pubic, and axillary hair), short stature and low body weight, rough skin and poor appetite (Figure 10). Inasmuch as zinc is an important constituent of erythrocytes, it appeared possible that continued hemolysis for a long period of time in patients with sickle cell disease might lead to a zinc deficient state, which could account for some of the clinical manifestations mentioned above. Delayed healing of leg ulcers and the reported beneficial effect of zinc therapy on leg ulcers in

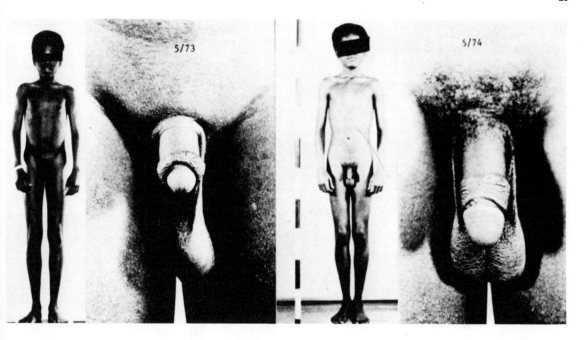

FIGURE 10.   A sickle cell anemia patient before and after zinc therapy. Changes in external genitalia and growth of pubic hair as result of therapy are evident. Height at 5/73, 144 cm; 5/74, 149 cm. (From Prasad, A. S., Schoomaker, E. B., Ortega, J., Brewer, G. J., Oberleas, D., and Oelshlegel, F. J., *Clin. Chem.*, 21, 582, 1975. With permission.)

sickle cell anemia patients also appears to be consistent with the above hypothesis.

In a study reported by Prasad et al.[120] zinc in plasma, erythrocytes, and hair was decreased and urinary zinc excretion was increased in sickle cell anemia patients as compared to controls (Table 12). Erythrocyte zinc and daily urinary zinc excretion were inversely correlated in the anemia patients ($r = -0.71$, $p < 0.05$), suggesting that hyperzincuria may have caused zinc deficiency in these patients (Figure 11). Carbonic anhydrase, a zinc metalloenzyme, correlated significantly with erythrocyte zinc ($r = +0.94$, $p < 0.001$) (Figure 12). Plasma RNase activity was significantly greater in anemic subjects than in controls; this is consistent with the hypothesis that sickle cell anemia patients were zinc deficient. Zinc sulfate (660 mg/day) was administered orally to seven men and two women with sickle cell disease. Two 17-year-old males gained 10 cm in height during 18 months of therapy. All but one patient gained weight (Figure 13). Five of the males showed increased growth of pubic, axillary, facial, and body hair. In one, a leg ulcer healed in 6 weeks on zinc, and in two other patients some benefit of zinc therapy on healing of ulcers was noted (Table 13).

In a recent study, the basal serum testosterone, dihydrotestosterone, and androstenedione levels in sickle cell anemia patients were found to be decreased as compared to the controls of similar age groups.[186a] Serum leutinizing hormone and follicular-stimulating hormone levels before and after stimulation with gonadotropin-releasing hormone were consistent with primary testicular failure due to zinc deficiency in sickle cell anemia subjects.

These data show that some sickle cell anemia patients are zinc deficient, although the severity varied considerably from one patient to another in this group. In spite of the tissue zinc depletion in sickle cell anemia patients, the mean urinary excretion of zinc was higher than in the controls. This may have been a direct result of increased filtration of zinc by the glomeruli (owing to continued hemolysis), or there may have been a defect in tubular reabsorption of zinc somehow related to sickle cell anemia — a possibility that cannot be excluded at present. Continued hyperzincuria may have been responsible for tissue depletion of zinc, as suggested by a significant negative correlation between values for 24-hr urinary zinc excretion and erythrocyte zinc. At this stage, however, one cannot rule out additional

TABLE 12

Growth of Body Hair in Seven Sickle Cell Anemia Male Patients on Zinc Therapy[a]

Number	Pubic	Axillary	Facial	Abdominal	Chest	Zn therapy (weeks)
1	0	0	0	0	0	49
	3+	1+	0	0	0	
2	2+	0	0	0	0	42
	3+	2+	1+	0	0	
3	3+	2+	1+	0	0	60
	4+	3+	3+	3+	3+	
4	3+	2+	0	0	0	10
	3+	3+	2+	2+	1+	
5	3+	2+	1+	0	0	7
	3+	2+	1+	2+	1+	
6	3+	2+	1+	0	0	43
	3+	2+	1+	0	0	
7	3+	2+	0	1	0	4
	3+	2+	0	1	0	

Note: 0, none; 1+, fine hair; 2+, definite; 3+, definite and bushy; 4+, adult-like.

[a]Top line of data in each set is for pretreatment, bottom line is after the indicated number of weeks of therapy.

From Prasad, A. S., Schoomaker, E. B., Ortega, J., Brewer, G. J., Oberleas, D., and Oelshlegel, F. J., *Clin. Chem.*, 21(4), 585, 1975. With permission.

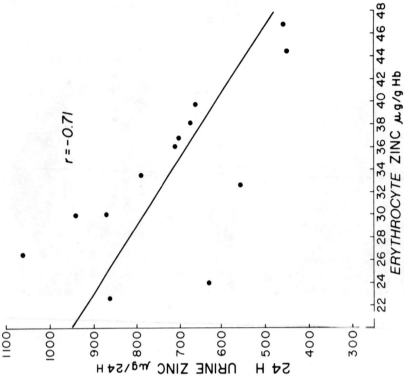

FIGURE 11. A 24-hr urinary zinc excretion compared with erythrocyte zinc content of sickle cell disease patients. (From Prasad, A. S., Schoomaker, E. B., Ortega, J., Brewer, G. J., Oberleas, D., and Oelshlegel, F. J., *Clin. Chem.*, 21, 582, 1975. With permission.)

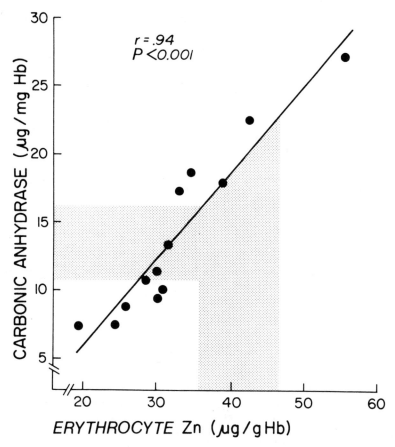

FIGURE 12. Carbonic anhydrase protein and zinc content of erythrocytes in sickle cell disease patients. Shaded areas indicate mean ± SD for erythrocyte zinc (vertical) and erythrocydle carbonic anhydrase protein (horizontal). From Prasad, A. S., Schoomaker, E. B., Ortega, J., Brewer, G. J., Oberleas, D., and Oelshlegel, F. J., *Clin. Chem.*, 21, 582, 1975. With permission.)

factors such as predominant dietary use of cereal protein and other nutritional factors that adversely affect zinc availability, thus accounting for zinc deficiency. Further work is warranted for proper elucidation of the pathogenesis of zinc deficiency in sickle cell anemia.

Recent studies have demonstrated a potential beneficial effect of zinc on the sickling process, in vitro, mediated by its effect on the oxygen dissociation curve[187,188] and the erythrocyte membrane.[189] Zinc-bound hemoglobin has increased oxygen affinity, a normal Bohr effect, and a decreased Hill coefficient.[187] Theoretically, an increased oxygen affinity of S hemoglobin should be beneficial to sickle cell disease patients inasmuch as this hemoglobin sickles upon deoxygenation. It is not known whether this zinc effect on S hemoglobin can be achieved in vivo. It is conceivable that the zinc-binding residue could

be one involved with the α_1-β_1 (or α_1-β_2) contact points, which might result in decreasing the stability of the contact and favor the oxy form of hemoglobin.[187] More definitive studies, such as X-ray diffraction, are needed before the exact Zn-binding residue and oxygen-affinity mechanism are defined with certainty.

It has been proposed that:

1. Hemoglobin binding to the inside of the sickled red cell membrane is an important feature of the development of the so-called irreversibly sickled cell.

2. Calcium facilitates the binding of hemoglobin particles.

3. Binding of hemoglobin particles to the membrane increases its stiffness and rigidity (perhaps by cross-linking mechanisms).

4. Zinc protects against these changes by

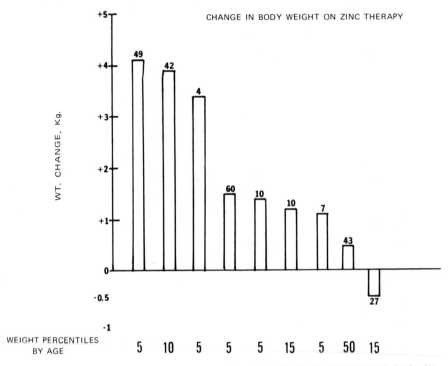

FIGURE 13. Changes in body weight of sickle cell anemia patients during zinc therapy. Number of weeks each patient had been on zinc therapy is given at top of columns. (From Prasad, A. S., Schoomaker, E. B., Ortega, J., Brewer, G. J., Oberleas, D., and Oelshlegel, F. J., *Clin. Chem.*, 21, 582, 1975. With permission.)

TABLE 13

Zinc in[a]

	Plasma (μg/dl)	Erythrocytes (μg/g Hb)	Hair (μg/g)	Urine (μg/g creatinine)	Plasma RNase ΔA/min/ml
Controls	112 ± 2.5 (23)[a]	41.7 ± 1.2 (23)	193 ± 4.3 (17)	495 ± 35 (10)	0.310 ± 0.001 (16)
Patients	102 ± 2.7 (32)	35.2 ± 1.6 (27)	149 ± 10.3 (21)	739 ± 60 (12)	0.435 ± 0.002 (31)
P	<0.025	<0.01	<0.01	<0.01	<0.001

Note: ±, SE.

[a]Numbers in parentheses are number of subjects. At least three 24-hr urines were collected from each subject.

From Prasad, A. S., Schoomaker, E. B., Ortega, J., Brewer, G. J., Oberleas, D., and Oelshlegel, F. J., *Clin. Chem.*, 21(4), 583, 1975. With permission.

partially blocking the binding of hemoglobin to the membrane either directly or by an interaction with calcium; as a result, less hemoglobin is bound to the membrane in the presence of zinc, and the deformability of the cell is thereby protected.[189]

In limited uncontrolled studies, zinc appears to have been effective in decreasing symptoms and crises of sickle cell anemia patients. The therapeutic rationale is based on effects of zinc on the red cell membrane by which it decreases hemoglobin and calcium binding and improves deformability, which may result in decreased trapping of sickle cells in the capillaries where pain cycle is normally initiated. Recent studies have shown that oral zinc therapy to patients with sickle cell anemia significantly lowers the number of circulating irreversibly sickle cells (ISCs).[189a] The role of ISCs in the pathology of sickle cell anemia is unknown, but some evidence suggests that ISCs may be related to vascular and organ damage, and as such zinc therapy may have the potential of reducing organ damage. Undoubtedly, a more thorough evaluation of this use of zinc therapy is needed in the future.

Acrodermatitis Enteropathica

Acrodermatitis enteropathica was described in 1943 by Danbolt and Closs,[190] and the clinical and pathological features have been delineated by numerous investigators.[191-197] In brief, acrodermatitis enteropathica is the expression of a lethal, autosomal, recessive trait which usually occurs in infants of Italian, Armenian, or Iranian lineage. The disease is not present at birth, but typically develops in the early months of life, soon after weaning from breast feeding. Dermatological manifestations include progressive bullous-pustular dermatitis of the extremities and the oral, anal, and genital areas, combined with paronychia and generalized alopecia. Infection with *Candida albicans* is a frequent complication. Ophthalmic signs may include blepharitis, conjunctivitis, photophobia, and corneal opacities. Gastrointestinal disturbances are usually severe, including chronic diarrhea, malabsorption, steatorrhea, and lactose intolerance. Neuropsychiatric signs include irritability, emotional disorders, tremor, and occasional cerebellar ataxia. The patients generally have retarded growth and hypogonadism. Prior to the discovery of diiodohydroxyquinolone therapy in 1953 by Dillaha, Lorincz, and Aavick,[198] patients with acrodermatitis enteropathica invariably died from cachexia, usually with terminal respiratory infection. Although diiodohydroxyquinolone has been used successfully for the therapy of this condition for 20 years, the mechanism of drug action has never been elucidated. It now seems possible that the efficacy of diiodohydroxyquinolone might be related to the formation of an absorbable zinc chelate, inasmuch as diiodohydroxyquinolone is a derivative of 8-hydroxyquinolone (a chelating agent).[199]

In 1973, Barnes and Moynahan[57,200] studied a 2-year-old girl with severe acrodermatitis enteropathica who was being treated with diiodohydroxyquinolone and a lactose-deficient synthetic diet. The clinical response to this therapy was not satisfactory, and the physicians sought to identify contributory factors. It was noted that the concentration of zinc in the patient's serum was profoundly reduced; therefore, they administered oral $ZnSO_4$. The skin lesions and gastrointestinal symptoms cleared completely, and the patient was discharged from the hospital. When $ZnSO_4$ was inadvertently omitted from the child's regimen, she suffered a relapse which promptly responded when oral $ZnSO_4$ was reinstituted. In their initial reports, Barnes and Moynahan[57,200] attributed zinc deficiency in this patient to the synthetic diet.

It was later appreciated that zinc might be fundamental to the pathogenesis of this rare inherited disorder and that the clinical improvement reflected improvement in zinc status. Support for the zinc deficiency hypothesis came from the observation that there was a close resemblance between the symptoms of zinc deficiency in animals and man, as reported earlier,[72] and subjects with acrodermatitis enteropathica, particularly with respect to skin lesions, growth pattern, and gastrointestinal symptoms.

Zinc supplementation to these patients led to complete clearance of skin lesions and restoration of normal bowel function, which had previously resisted various dietary and drug regimens. This original observation was quickly confirmed in other cases with equally good results.[73] The underlying mechanism of the zinc deficiency in these patients is, most likely, due to malabsorption. The cause of poor absorption is obscure, but an abnormality of Paneth's cells may be involved.

Miscellaneous Genetic Disorders

Low levels of plasma zinc have been noted in patients with mongolism,[55] the mechanism is unknown. Congenital hypoplasia of the thymus gland in cattle may be an example of zinc deficiency on a genetic basis.[201] It is not known whether thymus hypoplasia in man may be related to zinc deficiency.

Recently, an example of familial hyperzincemia has been reported.[201a] In five out of seven members of one family and two out of three second-generation individuals, the plasma zinc ranged from 250 to 435 μg %. Zinc levels in the red cells, hair, and bone were unremarkable, and no apparent clinical effects of hyperzincemia were observed.

Zinc Therapy in Rheumatoid Arthritis

Beneficial effects of oral zinc in therapeutic doses (zinc sulfate 220 mg three times daily) to patients with chronic, refractory rheumatoid arthritis have been reported recently.[201b] In a double-blind study, the zinc-treated patients did better than controls with respect to joint swelling and morning stiffness. It has not been established whether these subjects were deficient in zinc prior to therapy. The mode of action of zinc in this disease deserves further investigation.

METABOLIC ASPECTS OF ZINC IN HUMAN NUTRITION

Distribution in the Body

Several studies have been published on the distribution of zinc in tissues. Extensive analyses for trace elements in human tissues by emission spectrophotography were carried out by Tipton and Cook[202] and Tipton et al.[202a,202b] Approximately half of the subjects died from accidental causes, but the remainder died of various diseases. Other investigators have reported similar data, these are summarized in Table 14. For comparison, zinc levels of some animal tissues are also included in this table.

Liver, kidney, bone, retina, prostate, and muscle appear to be rich in zinc. In man, zinc content of testes and skin has not been determined accurately, although clinically it appears that these tissues are sensitive to zinc depletion. Hair is an easily accessible tissue. Zinc content of hair reflects more or less chronic nutriture. In our laboratory, the hair zinc level in normal subjects is 193 ± 18 μg/g.[203] Acute changes in zinc content of the body due to nutritional or other factors will not be reflected in hair zinc assays.

Zinc in Plasma and Red Cells

Zinc content in serum is 16% higher than in plasma.[204] The higher zinc content in serum has been attributed to the liberation of zinc from the platelets during the process of clotting and to invisible hemolysis of red cells, which occurs

TABLE 14

Zinc Concentrations in Human and Animal Tissues (mg/kg dry weight)[a]

	Human	Rat Normal	Rat Zinc deficient	Calf Normal	Calf Zinc deficient	Pig Normal	Pig Zinc deficient
Liver	141−245	101 ± 13	89 ± 12	101	84	150.8 ± 12	96.1 ± 8
Kidney	184−230	91 ± 3	80 ± 3	73	76	97.8 ± 3.0	90.8 ± 4.0
Lung	67−86	81 ± 3	77 ± 9	81	72		
Muscle	197−226	45 ± 5	31 ± 6	86	78		
Pancreas	115−135					139.5 ± 4.0	88.3 ± 4.0
Heart	100	73 ± 16	67 ± 9				
Bone	218	168 ± 8	69 ± 6	78	63	95 ± 1.8	47 ± 1.6
Prostate							
Normal	520						
Hyperplasia	2330						
Cancer	285						
Eye							
Retina	571						
Choroid	562						
Ciliary body	288						
Testis		176 ± 12	132 ± 16	79	70	54 ± 2.0	59 ± 2.0
Esophagus		108 ± 17	88 ± 10			88.1 ± 3.0	97.6 ± 5.0

[a]Mean ± SD except human data which are expressed as distribution of published mean values.

From Halsted, J. A., Smith, J. C., Jr., and Irwin, M. I., *J. Nutr.*, 104, 345, 1974. With permission.

regularly. With few exceptions, values for the plasma zinc in normal subjects are in reasonably good agreement, although obtained by different investigators using various techniques. Better methods for avoiding contamination and more precise analytical tools now provide accurate data for plasma zinc. According to our techniques, plasma zinc concentration (mean ± SD) in normal subjects is 112 ± 12 μg%.[205]

Binding of zinc to amino acids and serum protein was studied in vitro by Prasad and Oberleas.[206] Following incubation of zinc-65 with pooled native human serum in vitro, ultrafiltrable zinc was determined to comprise 2 to 8% of the total serum zinc, while the zinc-albumin molar ratio was varied from 0.33 to 2.5. Under similar conditions, 0.2 to 1.2% of zinc was ultrafiltrable when predialyzed serum was used. In physiological concentrations, addition of amino acids to predialyzed serum increased ultrafiltrable zinc-65 several fold (Figures 14 and 15). Histidine, glutamine, threonine, cystine, and lysine showed the most marked effects in this regard (Figures 16 and 17). It is suggested that an amino acid-bound fraction of zinc may play an important role in the biological transport of this element. By means of

starch-block electrophoresis of predialyzed serum, the stable zinc content was determined to be highest in the albumin fraction, although smaller concentrations of zinc were found in the α-, β-, and γ-globulins as well (Figure 18). However, results obtained by using zinc-65-incubated predialyzed serum indicated a difference in the behavior of exogenous zinc as compared with the endogenous zinc bound to various serum proteins. In vitro studies using predialyzed albumin, haptoglobin, ceruloplasmin, α_2-macroglobin, transferrin, and IgG incubated with zinc-65 revealed that zinc was bound to all of the above proteins and that the binding of zinc to IgG was electrostatic in nature. Whereas amino acids competed effectively with albumin, haptoglobin, transferrin, and IgG for binding of zinc, a similar phenomenon was not observed with respect to ceruloplasmin and α_2-macroglobulin, suggesting that the latter two proteins exhibited a special binding property for zinc (Figure 19).

Zinc in the red blood cells has been measured by only a few investigators. The reported values in the literature vary from 10 to 14 μg/ml of red cells.[72,126,207,208] Variations and lack of standardization of techniques undoubtedly are

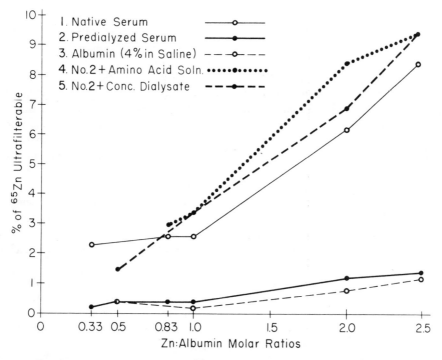

FIGURE 14.   Percent of ultrafiltrable [65]Zn plotted against zinc:albumin molar ratios. (From Prasad, A. S. and Oberleas, D., *J. Lab. Clin. Med.*, 76, 416, 1970. With permission.)

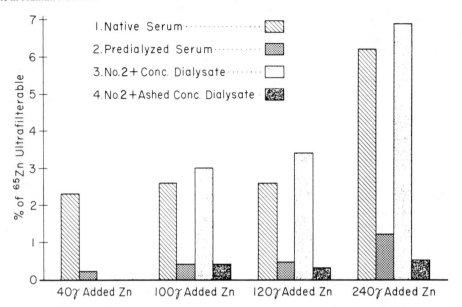

FIGURE 15.   Percent of ultrafiltrable ^{65}Zn in native serum, predialyzed serum, predialyzed serum reconstituted with concentrated dialysate, and predialyzed serum reconstituted with ashed concentrated dialysate at different levels of Zn:protein molar ratios (0.33, 0.83, 1.0, and 2.0). (From Prasad, A. S. and Oberleas, D., *J. Lab. Clin. Med.*, 76, 416, 1970. With permission.)

responsible for the widely different results. In our laboratory, normal values (mean ± SD) for red cell zinc are 42 ± 6 µg/g hemoglobin.[205] The expression of red cell zinc in terms of per gram of hemoglobin is preferable, inasmuch as accurate measurement of red cell volume is difficult. Although leukocytes are rich in zinc, only limited data were available for them in the literature; technical difficulties are mainly responsible for this. With further refinement in techniques, it is hoped that more can be learned from leukocyte zinc analysis in the future.

Absorption of Zinc

Normally, only a small percentage of ingested dietary zinc is absorbed. Absorption is difficult to precisely ascertain, because excretion of zinc is nearly all via the gut. Thus, intake-output studies indicating an increased absorption may be interpreted as showing a decreased excretion and vice versa.

Data on the site(s) of absorption in man and on the mechanism(s) of absorption (whether by active, passive, or facultative transport) are meager. Using the everted gut sac of the rat, Pearson and Reich,[209] provided evidence that zinc is actively absorbed from the distal gut segments against a concentration gradient. On the other hand, Methfessel and Spencer,[210] using zinc-65, studied specific absorption sites in rats by means of ligated intestinal sacs and concluded that the absorption of zinc-65 was significantly greater from the duodenum than from the more distal segments of the small intestines. Their data suggest that sites of zinc absorption may be similar to those of iron.

Becker and Hoekstra[211] concluded that "zinc absorption is variable in extent and is highly dependent upon a variety of factors." The factors which they suggested might affect zinc absorption include body size; level of zinc in diet; and presence in the diet of other potentially interfering substances such as calcium, phytate, other chelating agents, and vitamin D.

Evans[117] proposed the following mechanism for zinc absorption from the intestine: the pancreas secretes a ligand into the duodenum, where zinc complexes with the molecule; complexed with a ligand, zinc is transported through the microvillus and into the epithelial cell where the metal is transferred to binding sites on the basolateral plasma membrane; metal-free albumin interacts with the plasma membrane and removes zinc from the receptor sites. The quantity of metal-free albumin available at the basolateral membrane probably determines the amount of

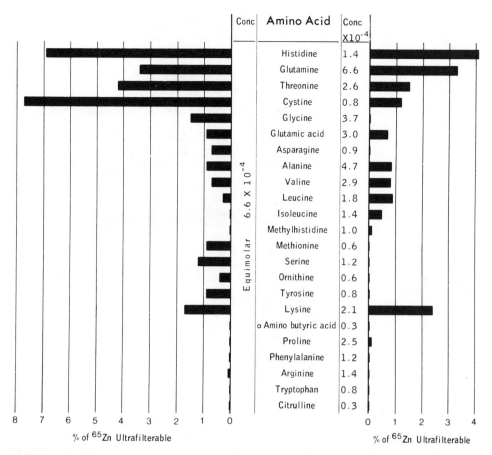

FIGURE 16. Effect of addition of single amino acids to predialyzed serum on the percent of ultrafiltrable ^{65}Zn:albumin ratio of 2.0 concentration of amino acids is expressed as $\times 10^{-4}$ *M*. (From Prasad, A. S. and Oberleas, D., *J. Lab. Clin. Med.*, 76, 416, 1970. With permission.)

zinc removed from the intestinal epithelial cell and thus regulates the absorption of zinc. This hypothesis must be regarded as speculative because much work is still needed in the future to substantiate it.

Availability of Zinc

Several factors influence the absorption and retention of zinc and thus its availability from the diet. Phytate (inositol hexaphosphate), which is present in cereal grains, markedly impairs the absorption of zinc. This was first shown in 1960 by O'Dell and Savage.[212] Later Oberleas, Muhrer, and O'Dell[37] showed that phytic acid added to an animal protein diet depressed growth in swine. A close relationship between zinc and the utilization of soybean protein was demonstrated in rats by Oberleas and Prasad.[213] Without zinc supplementation, rats fed a 12% soybean protein diet gained less than half as much as rats that were supplemented with zinc. Likuski and Forbes[214] showed

that phytic acid depressed the availability of zinc whether the protein source was pure amino acids or casein.

Phytate may exert a similar effect on zinc availability in man. Reinhold et al.[83] reported that unleavened bread, consumed in large amounts by the Iranian villagers (often providing the major source of protein), contains significantly more phytate than urban breads which are leavened. Leavening results in destruction of phytate. The omission of the leavening process in Iranian village bread-making is presumably responsible for the high content of phytate. Human zinc deficiency was reported by Prasad et al.[18] to occur in the Middle East under conditions where unleavened bread was consumed in great amounts, thus supporting the suggestion that phytate enhances the possibility of zinc deficiency in man.

Recent studies indicate that high fiber intake, which is common in subjects consuming high cereal proteins, is detrimental to zinc availa-

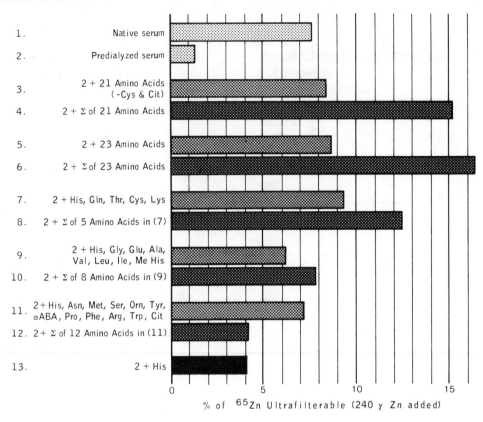

1. Native serum
2. Predialyzed serum
3. 2 + 21 Amino Acids (-Cys & Cit)
4. 2 + Σ of 21 Amino Acids
5. 2 + 23 Amino Acids
6. 2 + Σ of 23 Amino Acids
7. 2 + His, Gln, Thr, Cys, Lys
8. 2 + Σ of 5 Amino Acids in (7)
9. 2 + His, Gly, Glu, Ala, Val, Leu, Ile, Me His
10. 2 + Σ of 8 Amino Acids in (9)
11. 2 + His, Asn, Met, Ser, Orn, Tyr, αABA, Pro, Phe, Arg, Trp, Cit
12. 2 + Σ of 12 Amino Acids in (11)
13. 2 + His

% of ⁶⁵Zn Ultrafilterable (240 γ Zn added)

FIGURE 17.   Percent of ^{65}Zn ultrafilterable (240 γ Zn added). Effects of additions of physiological concentrations of amino acids in various combinations to samples of predialyzed serum, compared to the sum of their individual effects. Zn:albumin molar ratio = 2.0; Σ, sum of individual effects; Cys, cystine; Cit, citrulline; His, histidine; Gln, glutamine; Thr, threonine; Lys, lysine; Gly, glycine; Glu, glutamic acid; Ala, alanine; Val, valine; Leu, leucine; Ile, Isoleucine; Me His, 1-methyl-histidine; Asn, asparagine; Met, methionine; Ser, serine; Orn, ornithine; Tyr, tyrosine; αABA, α-aminobutyric acid; Pro, proline; Phe, phenylalanine; Arg, arginine; Try, tryptophan. (From Prasad, A. S. and Oberleas, D., *J. Lab. Clin. Med.,* 76, 416, 1970. With permission.)

bility.[83] Binding of zinc to the fiber of wheat is particularly important because, in contrast to other components, fiber is not degraded by digestive secretions. As a result, zinc remains attached to the fiber and is transported in this state into the large intestine where absorption does not occur. Ultimately, it is lost in the feces.

In vitro experiments suggest that binding of zinc and iron by phytate may be of secondary importance to binding by fiber.[83] This conclusion needs to be confirmed by studies of digestion and additional balance studies in man and other species of animals. Calcium enhances the phytate binding of zinc; however, its adjunctive behavior in the presence of fiber needs to be clarified.

Geophagia

Minnich et al.[20] reported that clay from Turkey inhibited iron absorption in human subjects. In Iran nearly all subjects studied with zinc deficiency and dwarfism gave a history of eating large amounts of clay for many years. It was, therefore, logical to suspect that Iranian clay might also hinder zinc absorption. Indeed, when a solution of zinc-65 was mixed with clay, 97% of the radioactivity was removed from the solution. However, when Iranian clay was fed to zinc-deficient rats, it proved to be a life-saving source of zinc.[47] It is possible, therefore, that subjects in Iran may have sought zinc through the ingestion of clay. At present, however, no definite conclusions can be drawn with respect to the effect of geophagia on zinc balance in man.

Chelating Agents

The effect of chelating agents on zinc

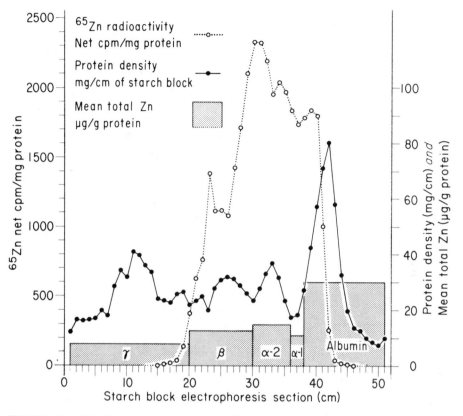

FIGURE 18. Distribution of endogenous stable zinc and exogenous ^{65}Zn starch block following electrophoresis of predialyzed normal human incubated with ^{65}Zn (without carrier zinc). (From Prasad, A. S. and Oberleas, D., *J. Lab. Clin. Med.*, 76, 416, 1970. With permission.)

absorption has been studied in experimental animals.[74] Ethylenediaminetetraacetic acid (EDTA) readily complexes with zinc from dietary sources and makes itself available for absorption. Although in small amounts EDTA may be beneficial with respect to zinc availability, therapeutic use of penicillamine in one patient with Wilson's disease resulted in zinc deficiency.[170]

Intake and Excretion of Zinc in Man

Up to the recent past, meager information was available on man's zinc requirement. Zinc balance studies carried out nearly 30 years ago indicated that the zinc balance was negative in preschool children receiving the lower zinc intake on a dietary zinc intake ranging from 4 to 6 mg/day, while retention of zinc was reported to be as high as 3.4 mg/day in others receiving the higher intake.[215] A high retention of zinc was also reported in other studies conducted in children and young adults. In the age group 8 to 12 years, the mean retention of zinc was 4.9 mg/day on a

zinc intake level ranging from 14 to 18 mg/day.[216] This indicates that there is a high requirement for zinc during growth and development. In young persons ranging in age from 17 to 27 years, the retention of zinc was also reported to be very high (5 to 8 mg/day) on a zinc intake ranging from 12 to 14 mg/day.[217] In a study reported in 1966, about 30% of a dietary zinc intake averaging 7 mg/day was retained in preadolescent girls. Variable results have been reported for adults. In early studies the zinc balance was reported to be in equilibrium in three subjects on a zinc intake of 5 mg/day; however, the retention of zinc was similar in two other subjects receiving twice this amount, while the zinc balance was more positive (ranging up to 2.6 mg/day on a zinc intake of 20 mg/day).[219] In a long-term study of two adults receiving a self-selected diet, one subject was in negative zinc balance (-4 mg/day) on a diet containing 11 mg/day, while the other (receiving a dietary zinc intake of 18 mg/day) was in a positive zinc balance of 1 mg/day.[220] In a recent study

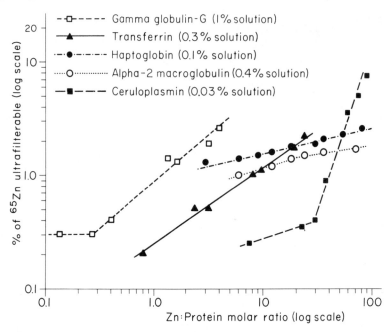

FIGURE 19.   Zn:protein molar ratio (log scale). Changes in percentages of ultrafiltrable [65]Zn at different levels of Zn:protein molar ratios are shown here for various predialyzed proteins. Data have been presented on a log-to-log scale. (From Prasad, A. S. and Oberleas, D., *J. Lab. Clin. Med.,* 76, 416, 1970. With permission.)

conducted in young women, the zinc balance was in equilibrium on a dietary zinc intake of about 11 mg/day,[221] while a study in young men receiving a synthetic diet with a zinc content of about 20 mg/day, the retention of zinc was high, ranging from 7 to 8 mg/day.[222]

The protein content of the diet appears to influence the body retention of zinc. In preadolescent girls receiving a low-protein diet and zinc intake of about 5 mg/day the zinc retention ranged from 0.5 to 0.8 mg/day,[223] while on a high-protein intake (with zinc intake of approximately 7 mg/day) the retention of zinc was about 2 mg/day. The obligatory zinc retention during the periods of growth and during stress has been calculated.[62] The zinc retention for growing children (aged 11 years) and for young adults (aged 17 years) was estimated to exceed 0.4 mg/day; for pregnant women the retention was 0.7 mg/day. The zinc content of school lunches in Maryland determined by emission spectroscopy was reported to average 3.9 mg for children ranging in age from 10 to 12 years.[224] The range of zinc content of these meals was 1.8 to 7.9 mg.

The zinc content of the diet greatly depends on the dietary protein content. A diet containing about 1 g protein per kilogram body weight for a 70-kg man is expected to contain about 12.5 mg zinc. A diet that is adequate in calories but has a low protein content may contain less than half this amount of zinc, whereas the zinc content of a high protein diet may be two to three times as high as that of the low protein diet (Table 15). The main source of the dietary zinc uptake is meat. However, marked differences have been demonstrated in zinc content in different types of meat; red meat has the highest zinc content.[125] Fish also has a relatively high zinc content. Most other food items which constitute the daily diet have a zinc content of 1 mg or less per 100 g net weight.

In one study, on a normal zinc intake of about 12.5 mg/day, the urinary zinc excretion ranged from 0.5 to 0.8 mg/day, while most of the zinc was excreted in stool.[76] The fecal excretion of zinc is only partly due to unabsorbed zinc, while the remainder is due to the secretion of zinc from the vascular space into the intestine. Intravenous studies of zinc-65 in man have shown that the intestinal secretion of zinc-65 may be as high as 18% of the administered zinc.[225,226] Zinc balances in adults were in equilibrium on a zinc intake of 12.5 mg/day. However, some studies

TABLE 15

Zinc Content of Meals of the General Hospital Diet

	Zinc content (mg/meal)	
Meal	Diet 1	Diet 2
Breakfast	1.1	3.6
Lunch	1.9	8.3
Supper	1.7	6.2
Total	4.7	18.1

Note: Calorie content: 2600 per day. Protein content: diet 1 = 69 g; diet 2 = 89 g.

From Spencer, H., Osis, D., Kramer, L., and Norris, C., in *Trace Elements in Human and Disease,* Vol. 1, Prasad, A. S., Ed., Academic Press, New York, 1976, 347. With permission.

TABLE 16

Zinc Balances During Different Dietary Zinc Intakes[a]

	Zinc (mg/day)	
Patient	Intake	Balance
1	11.6	−1.0
2	12.7	−0.9
3	15.0	+0.1
4	14.7	+0.2
5	14.6	+0.6

[a]The zinc intake was due to the dietary zinc content.

From Spencer, H., Osis, D., Kramer, L., and Norris, C., in *Trace Elements in Human Health and Disease,* Vol. 1, Prasad, A. S., Ed., Academic Press, 1976, 348. With permission.

TABLE 17

Zinc Balances During Different Intake Levels of Zinc (mg/day)

	Zinc (mg/day)		
Intake[a]	Urine	Stool	Balance
6.5	0.6	7.2	−1.3
12.2	0.5	11.0	+0.7
16.7	0.8	14.3	+1.6

[a]Changes in zinc intake are due to changes in protein intake.

From Spencer, H., Osis, D., Kramer, L., and Norris, C., in *Trace Elements in Human Health and Disease,* Vol. 1, Prasad, A. S., Ed., Academic Press, New York, 1976, 350. With permission.

have shown that not all individuals are in zinc equilibrium on this intake and that in some cases a higher zinc intake of about 15 mg/day is needed to attain equilibrium. Table 16 shows examples of zinc balances which were negative (about 1 mg/day) on a 12 mg zinc intake per day and zinc balances which were in equilibrium at a zinc intake of 14 to 15 mg/day. All zinc balance studies in man reported here must be considered as maximal

balances since the loss of zinc in sweat has not been considered and the balances are based only on the zinc intake and the excretion of zinc in urine and stool. In man the loss of zinc through sweat has been reported to be 1.15 mg/l for whole sweat and 0.9 mg/l for cell-free sweat.[227] If these amounts of zinc are lost in sweat, the requirement of 15 mg zinc per day is certainly not excessive.

Values for excretion of zinc and for the zinc balances during different intakes are shown in Table 17. On the lowest zinc intake of 6.5 mg/day the zinc balance was negative, −1.3 mg/day, while on the highest zinc intake (16.5 mg/day) the daily retention of zinc was about 2 mg/day.[76] In other studies carried out on a zinc intake of about 20 mg/day during a high protein intake, the zinc balance was in about the same range as that for the 16-mg intake. These data indicate that an insufficient protein intake and, therefore, an insufficient intake of zinc, if continued over a prolonged period of time, could lead to zinc deficiency, as would be expected to occur in conditions of protein malnutrition.

A zinc balance study was conducted in a 55-year-old male[76] when the intake of zinc was increased to 135 mg/day by adding zinc sulfate to a normal protein diet. The urinary zinc excretion increased from about 0.5 mg/day in the control study to about 3.5 mg/day during the high zinc intake (Figure 20). The increase of the urinary zinc excretion was greatest in the initial 24 days of the study period. Thereafter, the urinary zinc excretion reached a plateau and remained at this

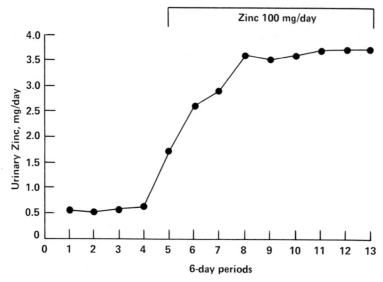

FIGURE 20.   Effect of zinc supplementation on urinary zinc excretion in man. A daily dose of 100 mg zinc sulfate was given orally. (From Spencer, H., Osis, D., Kramer, L., and Norris, C., in *Trace Elements in Human Health and Disease,* Vol. 1, Prasad, A. S., Ed., Academic Press, New York, 1976, 351. With permission.)

level for another 30 days. At the same time the plasma level of zinc increased gradually from 70 µg % to a maximum of 135 µg% in 45 days. Similar observations have been reported in other studies that were conducted during a normal protein intake in adults of normal body weight. In some of these studies the zinc balance increased by 2 mg/day, although the zinc intake was increased by 140 mg/day. In cases of malnutrition, the increase in zinc intake may result in a very high positive zinc balance.

During an intake of a low calorie diet used for weight reduction, the dietary protein content is usually low; therefore, the dietary zinc intake may be very low (about 2 mg/day). Figure 21 shows that under these conditions the loss of zinc may be high and the zinc balance may be negative (– 2.5 to 4 mg/day). In a study carried out during weight reduction in an obese patient using low calorie diets with different protein and zinc contents, a marked loss of zinc was demonstrated.[76] The loss of zinc was greatest during the lowest zinc intake (about 2.4 mg/day), and the loss of zinc decreased when the zinc intake was increased to 7.7 mg/day (Patient 1). Following the depletion period of 96 days, repletion with a high protein diet containing 25 mg zinc per day resulted in a retention of about 10 mg zinc per day. In a similar study carried out in Patient 2 after protein and zinc depletion, repletion with a normal high-calorie protein diet,

with a zinc content of about 35 mg/day, resulted in a zinc balance of 13.6 mg/day.

Individuals with a normal body weight who maintain themselves on a low calorie intake for prolonged periods of time (for example, patients with chronic alcoholism) show an increased retention of zinc when given a normal dietary intake. This high zinc retention was observed to continue for several months until the ideal body weight was attained.

When the dietary intake of zinc is completely eliminated, as in total starvation used for weight reduction in marked obesity, the rapid weight loss is associated with a considerable loss of zinc. This zinc loss occurs promptly with onset of the weight loss and is primarily due to a marked increase of the urinary zinc excretion.[76] Figure 22 shows that the daily urinary zinc excretion increased twofold in the first 6 days of starvation and that this increase persisted throughout the 60 days of starvation. The average zinc loss per day was 4.6 mg and the total zinc loss was 276 mg. Considering the assumed total body zinc of 2.5 g, the loss of zinc during weight reduction, induced either by starvation or by a low calorie diet, may be as great as 10 to 15% of the total body zinc.

During starvation the high excretion of zinc resulting in weight loss is most likely due to an increased muscle breakdown. Although muscle tissue has a very low concentration of

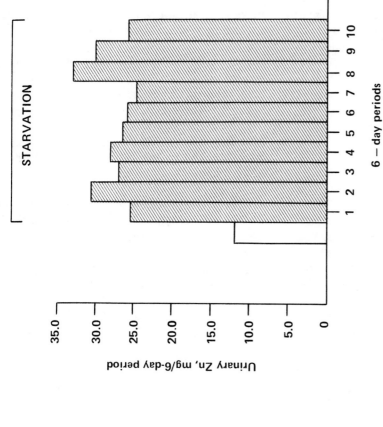

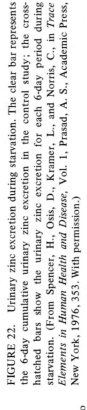

FIGURE 22. Urinary zinc excretion during starvation. The clear bar represents the 6-day cumulative urinary zinc excretion in the control study; the cross-hatched bars show the urinary zinc excretion for each 6-day period during starvation. (From Spencer, H., Osis, D., Kramer, L., and Norris, C., in *Trace Elements in Human Health and Disease*, Vol. 1, Prasad, A. S., Academic Press, New York, 1976, 353. With permission.)

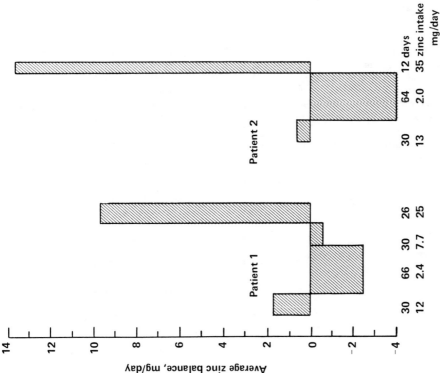

FIGURE 21. Zinc balances during different zinc intakes in man. Studies of two obese patients during weight reduction. In the control study the dietary zinc intake was about 12 mg/day. On a 600-cal intake the average dietary zinc intake was 2.2 mg/day, and in one study 7.7 mg/day (Patient 1). During the high protein 1800-cal refeeding period, the dietary zinc intakes were 25 mg and 35 mg/day, respectively. (From Spencer, H., Osis, D., Kramer, L., and Norris, C., in *Trace Elements in Human Health and Disease*, Vol. 1, Prasad, A. S., Academic Press, New York, 1976, 352. With permission.)

zinc,[126],[225],[226] the muscle mass is very large and contains a high percentage of the total body zinc.[228] During weight loss the loss of muscle tissue is considerable, in addition to the loss of fat and water. In studies of zinc metabolism following injury and surgical procedures, the increased excretion of zinc in urine has been ascribed to loss of muscle tissue.[229],[230] The source of the zinc loss was related to muscle tissue because of the good correlation between the urinary zinc/creatin ratio to the urinary excretion of nitrogen.[229]

BIOCHEMISTRY AND PHYSIOLOGY OF ZINC

Various studies by Vallee[231] have shown that zinc is a constituent of a number of metallo-enzymes. Although iron and copper enzymes have been recognized for some time (owing to their characteristic colors) zinc metalloenzymes have only recently emerged on the scene.

The first concrete demonstration of a specific biological function that was critically dependent on the presence of zinc came with the discovery of Keilin and Mann[232] showing that carbonic anhydrase contained zinc that was essential to its mechanism of action. Over the next 20 years only five additional zinc metalloenzymes were identified, but in the last 15 years the total number has risen to about 24 (Table 18). If related enzymes from different species are included, then over 70 zinc metalloenzymes are now on record.

Zinc metalloenzymes exhibit perhaps the greatest diversity of both catalytic function and the role played by the metal atom.[232-234] They are now known to be present throughout all phyla and to participate in a wide variety of metabolic processes including carbohydrate, lipid, protein, and nucleic acid synthesis or degradation. Each of the six categories of enzymes designated by the International Union of Biochemistry (IUB) Commission on Enzyme Nomenclature contains at least one example of a zinc metalloenzyme. The metal is present in several dehydrogenases, aldolases, peptidases, and phosphatases. Since the zinc cation has a d_{10} electronic configuration, it

TABLE 18

Zinc Metalloenzymes

Enzyme	IUB no.	Source
Alcohol dehydrogenase	1.1.1.1	Yeast; horse, human liver
D-Lactate cytochrome reductase	1.1.2.4	Yeast
Glyceraldehyde-phosphate dehydrogenase	1.2.1.13	Beef, pig muscle
Phosphoglucomutase	2.7.5.1	Yeast
RNA polymerase	2.7.7.6	*E. coli*
DNA polymerase	2.7.7.7	*E. coli*; sea urchin
Reverse transcriptase	2.7.7.–	Avian myeloblastosis virus
Mercaptopyruvate sulfur transferase	2.8.1.2	*E. coli*
Alkaline phosphatase	3.1.3.1	*E. coli*
Phospholipase C	3.1.4.3	*Bacillus* cereus
Leucine aminopeptidase	3.4.1.1	Pig kidney, lens
Carboxypeptidase A	3.4.2.1	Beef, human pancreas
Carboxypeptidase B	3.4.2.2	Beef, pig pancreas
Carboxypeptidase G	3.4.2.–	*Pseudomonas stutzeri*
Dipeptidase	3.4.3.–	Pig kidney
Neutral protease	3.4.4.–	*Bacillus* sp.
Alkaline protease	3.4.4.–	*Escherischia freundii*
AMP aminohydrolase	3.5.4.6	Rabbit muscle
Aldolase	4.1.2.13	Yeast; *Aspergillus niger*
Carbonic anhydrase	4.2.1.1	Erythrocytes
δ-Aminolevulinic acid dehydratase	4.2.1.24	Beef liver
Phosphomannose isomerase	5.3.1.8	Yeast
Pyruvate carboxylase	6.4.1.1	Yeast

From Riordan, J. F. and Vallee, B. L., in *Trace Elements in Human Health and Disease,* Vol. 1, Prasad, A. S., Ed., Academic Press, New York, 1976, 228.

invariably exists in the +2 oxidation state and does not undergo oxidation or reduction. The development of highly precise, rapid, and convenient means for zinc analysis no doubt accounts for much of the increase in the rate of recognition of zinc metalloenzymes.

Chemical stability is probably an essential aspect of the utilization of zinc in diverse biological processes such as hydrolysis, transfer, and addition to double bonds, and even oxidoreduction.[253] However, the role of zinc in redox enzymes such as alcohol dehydrogenase is not to donate or accept electrons; rather, it serves as a Lewis acid. It is this capacity to serve as a super acid that probably underlies the function of zinc in many zinc metalloenzymes. A zinc metalloenzyme is defined as a catalytically active metalloprotein containing stoichiometric amounts of zinc firmly bound at its active site.[235] The metal atoms are so tightly bound that they do not dissociate from the protein during the isolation procedure. When the metal is loosely bound, the association is chemically and functionally more tenuous, and the designation "metal-enzyme complex" has been employed to convey this distinction.

In zinc metalloenzymes the metal is located at the active site and participates in the actual catalytic process. However, this may not be the only function of zinc in enzymes. It may serve to stabilize structure, as in *Bacillus subtilis* α-amylase,[231] or it may have a regulatory role, as in aspartate transcarbamylase.[236] It can serve in both catalysis and structure, e.g., horse liver alcohol dehydrogenase,[237] or in both catalysis and regulation, e.g., bovine lens aminopeptidase.[238]

In the past few years zinc has been found in both DNA and RNA polymerases. The recent demonstration that RNA-dependent DNA polymerase in the reverse transcriptase of avian myeloblastosis and other viruses is also a zinc metalloenzyme indicates a relationship between zinc metabolism and malignancy. This opens up new approaches to the investigation of cancer. Recent studies in biological experiments indicate that thymidine kinase is also a zinc-dependent enzyme and that it is very sensitive to a lack of zinc.[239] The activity of RNase is also regulated by exogenous zinc; thus, zinc appears to play a very important role in RNA and DNA metabolism.

Zinc Enzymes in Zinc Deficiency

In view of the fact that zinc is needed for many enzymes, one may speculate that the level of zinc in cells controls the physiological processes through the formation and/or regulation of activity of zinc-dependent enzymes. However, until 1965 there was no evidence in the literature to support this concept.[72] Since then, studies have been reported showing that the activity of various zinc-dependent enzymes (as judged by histochemical techniques) was reduced in the testes, bones, esophagus, and kidneys of zinc-deficient rats in comparison to their pair-fed controls.[72,240] These results correlated with the decreased zinc content in the above tissues of the zinc-deficient rats and the clinical manifestations of testicular atrophy, reduced growth rate, and esophageal parakeratosis. This suggests that the likelihood of detecting any biochemical changes is greatest in tissues that are sensitive to zinc depletion. This subject has recently been reviewed by Kirchgessner et al.[240a]

Within the past 8 years many reports have appeared describing changes in activities of the alkaline phosphatase that was found to be significantly decreased in the serum of zinc-deficient rats, pigs, dairy cows, calves, and chicks. Only in the study by Kfoury, Reinhold, and Simonian[241] was there no reduction in the activity of the serum alkaline phosphatase in the plasma of zinc-deficient rats. In other studies the serum or plasma activities were not different from those of zinc-supplemented, pair-fed control rats, but were significantly lower than those of ad libitum-fed controls,[242-244] indicating that a reduction in food intake and not deficiency of zinc per se was responsible for these results.

The alkaline phosphatase in serum responds rapidly to zinc depletion through a significant reduction of its activity. In the studies by Roth and Kirchgessner[245] this enzyme lost 27 and 48% of its activity within 2 and 4 days, respectively, after the feeding of the zinc-deficient diet was initiated. The activity in the zinc-deficient group was significantly lower than in pair-fed control rats, although there was no difference between the two zinc-supplemented control groups, as illustrated in Figure 23. This is in agreement with Guttikar; Panemangalore, and Roa[246] who found that the alkaline phosphatase was not affected by calorie intake. Since the activity of this enzyme

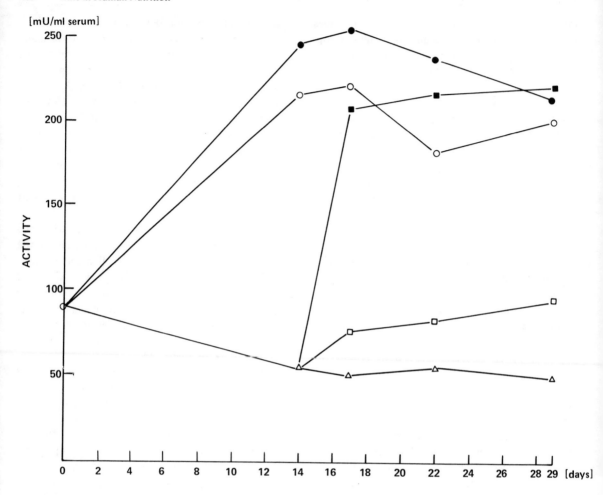

FIGURE 23.   Activities of alkaline phosphatase in serum of depleted, repleted, and control rats. (■) Zinc-repleted rats (12 ppm Zn); (●) pair-fed controls (96 ppm Zn); (○) control rats (96 ppm Zn); (□) zinc-repleted rats (4.5 ppm Zn); (△) zinc-deficient rats (1.2 ppm Zn). (From Kirchgessner, M. and Roth, H. P., *Arch. Tierernaehr.*, 25, 83, 1975. With permission.

decreased before any signs of a lowered food intake or reduced growth rate were evident, it can be concluded that the activity loss is directly attributed to zinc deficiency. As shown in Figure 23, the activity of the serum alkaline phosphatase returned to the level of the control animals within 3 days after the deficient diet was supplemented with zinc.

In zinc-deficient subjects, as reported from the Middle East by Prasad, Halsted, and Nadimi[18] and Sandstead et al.,[23] serum alkaline phosphatase activity consistently increased following supplementation with zinc. Thus, it appears that an induction of serum alkaline phosphatase activity following zinc supplementation may indicate a zinc-deficient state in man.

In almost all studies,[240,241,243,251-253,259,260] the activity of the alkaline phosphatase was found to be reduced in bones from zinc-deficient rats, pigs, cows, chicks, turkey poults, and quail. In disagreement with this are the early report by Day and McCollum[247] and the studies by Shrader and Hurley,[248] who did not detect a reduced activity in the bone marrow of depleted rats.

Compared to serum, however, the alkaline phosphatase in the rat femur shows a lower turnover rate.[249] After 4 days of depletion, zinc-depleted rats had 21% lower activity in the bone (in serum, 48%) and 54% lower activity (in serum, 75%) in 30 days than ad libitum-fed control animals. The total zinc content of the femur also decreased by 54%, analogous to the loss in the activity of the alkaline phosphatase.[250] In the bone the activity of this enzyme is somewhat influenced by a restricted food intake and the retarded growth of the animals (see Figure 24). This is also true for turkey poults, as was found by Starcher and Kratzer.[251]

Alkaline Phosphatase in Intestinal and Other Tissues

The activity of alkaline phosphatase may be reduced in the intestine, kidneys and stomach in experimental animals due to zinc deficiency. There may not only be a loss of activity due to a lack of sufficient zinc for maintaining the enzyme activity, but the amount of enzyme present may also be diminished because of either a lowered synthesis or an increased degradation. This was suggested by Kfoury, Reinhold, and Simonian[241] and Luecke, Olman and Baltzer,[244] who studied intestinal mucosa, and by Davies and Motzok,[252] who were unable to raise the alkaline phosphatase activity in the tibia of chicks by preincubation with zinc to the level of the zinc-supplemented controls. The study by Iqbal,[253] using pair-fed

animals, indicates that the lowered activity of the intestinal alkaline phosphatase is probably not caused by a reduced food intake. Further support of this view is provided by Williams,[254] who noted a lower activity of this enzyme in intestinal tissue as early as 3 days after rats were given a zinc-deficient diet.

Pancreatic Carboxypeptidases

Two important enzymes in protein digestion are the pancreatic carboxypeptidases A and B. Table 19 shows a summary of changes in these zinc metalloenzymes of the pancreas in response to zinc deficiency. A loss of activity of the pancreatic carboxypeptidase A in zinc deficiency is a consistent finding. According to studies by Roth and Kirchgessner,[255] this enzyme lost 24% of its activity in the rat pancreas within 2 days of a dietary zinc depletion. Within 3 days zinc repletion rapidly restored the activity of the carboxypeptidase A to the normal levels of pair-fed animals (Figure 25). The level of food intake had no influence.

With regard to carboxypeptidase B, Hsu, Anilane, and Scanlan[256] did not find reduced

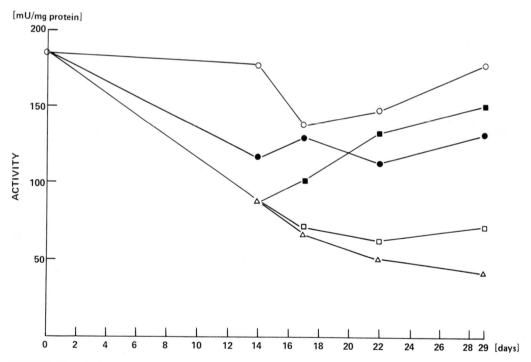

FIGURE 24. Activities of alkaline phosphatase in femora of depleted, repleted, and control rats. (■) Zinc-repleted rats (12 ppm Zn); (●) pair-fed controls (96 ppm Zn); (○) control rats (96 ppm Zn); (□) zinc-repleted rats (4.5 ppm Zn); (△) zinc-deficient rats (1.2 ppm Zn). From Kirchgessner, M. and Roth, H. P., *Arch. Tierernaehr.*, 25, 83, 1975. With permission.)

TABLE 19

Activity of the Carboxypeptidase A and B in Pancreas of Zinc-deficient Animals

Animal	Tissue	Enzyme activity[a]	Ref.
Pancreas carboxypeptidase A			
Rat	Pancreas	−	256
		−	281
		−	240
		−	255
Swine	Pancreas	−	262
Pancreas carboxypeptidase B			
Rat	Pancreas	0	256
		−	255

[a]0, unchanged; −, reduced

Modified from Kirchgessner, M., Roth, H. P., and Weigand, E., in *Trace Elements in Human Health and Disease,* Vol. 1, Prasad, A. S., Ed., Academic Press, New York, 1976, 204. With permission.

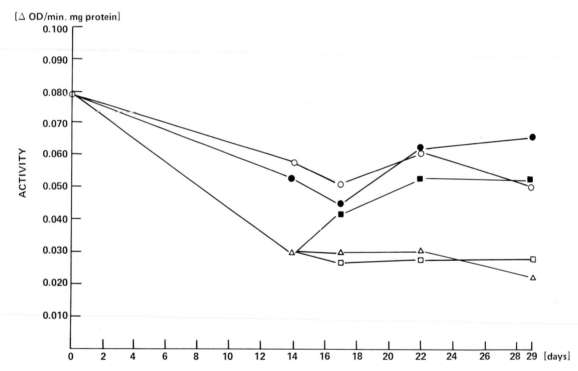

FIGURE 25. Activities of pancreatic carboxypeptidase A of depleted, repleted, and control rats. (■) Zinc-repleted rats (12 ppm Zn); (●) pair-fed controls (96 ppm Zn); (○) control rats (96 ppm Zn); (□) zinc-repleted rats (4.5 ppm Zn); (△) zinc-deficient rats (1.2 ppm Zn). (From Kirchgessner, M. and Roth, H. P., *Arch. Tierernaehr.,* 25, 83, 1975. With permission.)

activities in the zinc-deficient pancreas. However, Roth and Kirchgessner,[255] reported a loss of activity for this pancreatic enzyme by about 50% in zinc-depleted rats, compared to pair-fed and ad libitum-fed control animals.

Carbonic Anhydrase

Reduced activities have been observed.[240a] in tissues such as blood, stomach, and intestine (in which this enzyme has a major functional role). However, the decreased activities, which were found in the blood of rats,[257] calves,[14] and lambs[258] are in contrast to a number of studies that did not report such changes. In the studies by Roth and Kirchgessner[257] the activity of carbonic anhydrase was reduced by 41% after 4 days of zinc depletion. However, after 30 days there was no longer any difference between the deficient and the zinc-supplemented animals. At the same time, the erythrocyte count (expressed per cubic millimeters) of blood had increased by 41% in the deficient group. Therefore, when the enzyme activity was expressed per unit of erythrocytes, a reduction in the activity of the carbonic anhydrase of blood could be demonstrated at the beginning of the dietary zinc depletion and later during stages of severe zinc deficiency. Iqbal[253] found that zinc deficiency in rats reduced the carbonic anhydrase activity of gastric and intestinal tissue by 47 and 33%, respectively.

Recently in sickle cell disease patients, an example of a conditioned zinc-deficient state, the content of carbonic anhydrase in the red cells was found to be decreased, correlating with the zinc content of the red cells.[120] Inasmuch as the technique measured the apoenzyme content, it appears that zinc may have a specific effect on the synthesis of this protein. The precise mechanism by which zinc may effect the synthesis of the apoenzyme is not presently understood.

Alcohol Dehydrogenase

Primarily, the studies by Prasad et al.[259-262] and Prasad and Oberleas[240] clearly showed that zinc deficiency lowers the activity of this enzyme in the liver, bones, testes, kidneys, and esophagus of rats and pigs (Tables 20 and 21). Roth and Kirchgessner[263] found 28% lower activity of the alcohol dehydrogenase in the livers of severely zinc-deficient rats. Similar observations have also been reported by Kfoury, Reinhold, and Simonian.[241]

In a recent study, alcohol dehydrogenase was assayed in subcellular fractions of liver and retina from zinc-deficient and control rats using retinol and ethanol as substrates.[264] The activity of alcohol dehydrogenase was significantly decreased as a result of zinc deficiency. In older rats, although no changes in liver zinc and activity of alcohol dehydrogenase were found, the retina was shown to be sensitive to the lack of zinc. These data show that zinc is required for the metabolism

of vitamin A as well as the catabolism of ethanol. However, other investigators did not observe reduced activities in the liver[242] or the femur muscle of zinc-deficient rats.[265]

Glutamic Dehydrogenase

Glutamic dehydrogenase evidently is a zinc metalloenzyme with a high affinity for zinc. Apart from the study by Kfoury, Reinhold, and Simonian,[241] attempts to demonstrate a lowered activity of this enzyme in tissues of rats and pigs have not met with success.[240a] Roth and Kirchgessner[263] found that the activity of the glutamic dehydrogenase was even higher in the liver of zinc-deficient rats than in ad libitum-fed control animals; however, a comparable increase in activity was also observed in the liver of pair-fed controls.

Lactic and Malic Dehydrogenase

In most investigations using rats and pigs, the lactic and malic dehydrogenases were not found to decrease in response to zinc depletion, not even in the serum, pancreas, and bones, which are tissues known to be sensitive to zinc supply.[240a] Prasad et al.,[259,260] applying histological techniques, did observe lower tissue activities for lactic and malic dehydrogenase in zinc-deficient rats and pigs. However, these authors could only partially confirm their results by spectrophotometric analyses of the enzyme activities.[240,263] Similarly, reduced activities of these dehydrogenases were not detectable in the testes of zinc-deficient rats by either histological methods or spectrophotometric assay.[266] Even in the most severe stage of deficiency by dietary zinc depletion for 30 days, no loss of activity was found for both the lactic and malic dehydrogenase in serum and liver or for the lactic dehydrogenase in muscular tissue.[245,263,265] Compared to ad libitum-fed control rats, however, the liver lactic dehydrogenase had decreased by more than 50% in these studies. Since the pair-fed controls exhibited the same drop in enzyme activity, these responses are not directly attributable to zinc deficiency but rather to differences in food intake (see Figure 26). This inference is supported by the reports of Hurley[267] and Swenerton and Hurley.[268]

In contrast to the liver of the pig, the liver of the rat is not a zinc-sensitive organ.[262] This is confirmed by the observation that the change in the

TABLE 20

Zinc and Specific Activities of Various Enzymes in Tissues of Controls and Zinc-deficient Rats

| Tissues | Zinc, μg/g dry wt | Activity of enzymes (ΔOD/min/mg protein) | | | | | Alkaline phosphatase (sigma U/mg protein) | CPD (μmol β-naphthol liberated/mg protein) |
		ADH	Aldolase	LDH	ICDH	SDH		
Liver								
A	91.7 ± 4.7	0.021 ± 0.0022	0.107 ± 0.0064		0.43 ± 0.02	0.151 ± 0.012		
B	107.3 ± 3.1	0.022 ± 0.0013	0.134 ± 0.0098		0.38 ± 0.03	0.152 ± 0.017		
C	87.6 ± 3.5	0.013 ± 0.0008	0.124 ± 0.0100		0.38 ± 0.02			
A vs. B[a]	$p < 0.025$	NS	$p < 0.05$		NS			
A vs. C	NS	$p < 0.001$	NS		NS			
B vs. C	$p < 0.001$	$p < 0.001$	NS		NS	NS		
Kidney								
A	93 ± 2.2	0.0031 ± 0.00025	0.20 ± 0.01	2.9 ± 0.27	0.57 ± 0.03	0.188 ± 0.0075	3.3 ± 0.22	
B	87 ± 0.8	0.0031 ± 0.00012	0.22 ± 0.01	2.67 ± 0.07	0.48 ± 0.03	0.187 ± 0.0036	4.5 ± 0.25	
C	80 ± 0.9	0.0014 ± 0.00013	0.21 ± 0.01	2.53 ± 0.04	0.49 ± 0.04		2.3 ± 0.17	
A vs. B	NS	NS	NS	NS	NS		$p < 0.005$	
A vs. C	$p < 0.001$	$p < 0.001$	NS	NS	NS		$p < 0.005$	
B vs. C	$p < 0.001$	$p < 0.001$	NS	NS	NS	NS	$p < 0.001$	
Testis								
A	191.0 ± 3.1	0.022 ± 0.0018	0.161 ± 0.0034	1.20 ± 0.08	0.063 ± 0.0067	0.128 ± 0.0048	2.75 ± 0.11	
B	197.1 ± 2.9	0.028 ± 0.0030	0.161 ± 0.0047	1.22 ± 0.09	0.046 ± 0.0052	0.127 ± 0.0054	2.80 ± 0.10	
C	114.7 ± 4.9	0.012 ± 0.0014	0.164 ± 0.0112		0.090 ± 0.0076		1.70 ± 0.10	
A vs. B	NS	NS	NS		NS		NS	
A vs. C	$p < 0.001$	$p < 0.001$	NS		$p < 0.025$		$p < 0.001$	
B vs. C	$p < 0.001$	$p < 0.001$	NS	NS	$p < 0.001$	NS	$p < 0.001$	
Pancreas								
A	79.5 ± 3.2				0.18 ± 0.02		1.8 ± 0.18	64.8 ± 11.2
B	88.9 ± 3.7				0.16 ± 0.02		1.7 ± 0.11	61.19 ± 10.29
C	74.4 ± 3.2				0.14 ± 0.01		1.4 ± 0.05	10.69 ± 3.41
A vs. B	NS				NS		NS	NS
A vs. C	NS				NS		$p < 0.05$	$p < 0.001$
B vs. C	$p < 0.01$				NS		$p < 0.025$	$p < 0.001$
Bone								
A	180 ± 7.7	0.0020 ± 0.0001	0.128 ± 0.009				3.46 ± 0.23	

[a]Comparison of means.

TABLE 20 (continued)

Zinc and Specific Activities of Various Enzymes in Tissues of Controls and Zinc-deficient Rats

Tissues	Zinc, μg/g dry wt	Activity of enzymes (ΔOD/min/mg protein)					Alkaline phosphatase (sigma U/mg protein)	CPD (μmol β-naphthol liberated/mg protein)
		ADH	Aldolase	LDH	ICDH	SDH		
B	211 ± 5.4	0.0026 ± 0.0001	0.068 ± 0.006	0.82 ± 0.04			4.69 ± 0.16	
C	139 ± 9.6	0.0009 ± 0.0001	0.078 ± 0.009	0.76 ± 0.04			2.69 ± 0.14	
A vs. B	$p < 0.01$	$p < 0.001$	$p < 0.001$				$p < 0.001$	
A vs. C	$p < 0.01$	$p < 0.001$	$p < 0.001$				$p < 0.025$	
B. vs. C	$p < 0.001$	$p < 0.001$	NS	NS			$p < 0.001$	
Thymus								
A	83.7 ± 1.9		0.079 ± 0.008				6.2 ± 0.54	
B	98.8 ± 3.5		0.080 ± 0.006	4.41 ± 0.34			5.7 ± 0.32	
C	81.1 ± 2.2		0.077 ± 0.005	3.78 ± 0.52			3.5 ± 0.25	
A vs. B	$p < 0.005$		NS				NS	
A vs. C	NS		NS				$p < 0.001$	
B vs. C	$p < 0.005$		NS	NS			$p < 0.001$	

Note: A, ad libitum-fed controls, B, pair-fed controls; C, zinc-deficient rats; NS, not significant (P > 0.05%; ADH, alcohol dehydrogenase; LDH, lactic dehydrogenase; ICDH, isocitric dehydrogenase; SDH, succinic dehydrogenase; CPD, carboxypeptidase; ±, mean SE.

From Prasad, A. S. and Oberleas, D., *J. Appl. Physiol.*, 31, 842, 1971.

TABLE 21

Zinc, DNA, RNA, and Protein Content and the Activities of Various Enzymes in Tissues of Pair-fed Controls and Zinc-deficient Animals

Tissue[a]	Zinc (μg/g of dry wt)	DNA (μg/mg of wet wt)	RNA (μg/μg of DNA)	Protein (mg/mg of DNA)	ADH Initial	ADH % change After addition of Zn[b]	ADH % change After addition of EDTA[c]	LDH	ALD	SDH	ICDH	AP (U/mg of DNA)	CPD Initial	CPD % change After addition of Zn[b]	CPD % change After addition of EDTA[c]
Liver															
A	150.8 ± 12	4.2 ± 0.42	3.5 ± 0.34	40 ± 2.9	6.4 ± 0.69	89 ± 2	79 ± 4	46 ± 4	5.7 ± 0.7	—	—	—	—	—	—
B	96.1 ± 8	5.1 ± 0.3	2.5 ± 0.18	35 ± 1.9	3.6 ± 0.39	95 ± 3	77 ± 1	50 ± 4	3.0 ± 0.2	—	—	—	—	—	—
p[d]	<0.005	NS	<0.05	NS	<0.005			NS	<0.005						
Kidney															
A	97.8 ± 2.6	3.7 ± 0.16	1.4 ± 0.06	23.4 ± 0.8	0.162 ± 0.009	91 ± 4	NA	—	2.54 ± 0.11	1.6 ± 0.17	6.9 ± 0.72	154.8 ± 16.4	—	—	—
B	87.0 ± 2.8	3.9 ± 0.20	1.1 ± 0.05	21.2 ± 1.0	0.123 ± 0.008	96 ± 8	NA	—	1.86 ± 0.09	1.7 ± 0.19	7.2 ± 0.19	80.2 ± 9.5	—	—	—
p[d]	<0.025	NS	<0.005	<0.05	<0.025				<0.001	NS	NS	<0.005			
Bone															
A	98.5 ± 2.08	0.53 ± 0.07	3.04 ± 0.26	57.5 ± 5	2.30 ± 0.35	60 ± 10	NA	81 ± 10	21 ± 3.4	—	—	2640 ± 420	—	—	—
B	48.1 ± 2.74	0.83 ± 0.13	1.81 ± 0.22	36.8 ± 5	0.64 ± 0.07	50 ± 24	NA	44 ± 7	12 ± 2.5	—	—	480 ± 120	—	—	—
p[d]	<0.001	<0.1 > 0.05	<0.005	<0.025	<0.001			<0.025	<0.1 > 0.05			<0.001			
Pancreas															
A	133.3 ± 6	4.7 ± 0.28	5.0 ± 0.15	25 ± 1.8	—	—	—	14 ± 1.6	—	—	—	—	2.8 ± 0.1	82 ± 7	40 ± 16
B	96.8 ± 4	4.0 ± 0.30	4.1 ± 0.1	22 ± 2.2	—	—	—	14 ± 1.0	—	—	—	—	1.3 ± 0.2	70 ± 12	35 ± 15
p[d]	<0.001	<0.2 > 0.1	<0.05	NS				NS					<0.001		

Note: ADH, alcohol dehydrogenase; ALD, adolase; SDH, succinic dehydrogenase; CPD, carboxypeptidase; LDH, lactic dehydrogenase; ICDH, isocitric dehydrogenase; AP, alkaline phosphatase; NS, not significant; NA, no activity; ±, mean SE.

[a] A, pair-fed controls; B, zinc-deficient animals.

[b] Final concentration of Zn: liver, 8×10^{-5} M; kidney, 3.3×10^{-5} M; bone, 1.7×10^{-5} M; pancreas, 1.8×10^{-5} M.

[c] Final concentration of EDTA: liver, 6.7×10^{-3} M; kidney, 3.3×10^{-3} M; bone, 6.7×10^{-3} M; and pancreas, 7.3×10^{-3} M.

[d] Comparison of means.

Modified from Prasad, A. S., Oberleas, D., Miller, E. R., and Luecke, R. W., *J. Lab. Clin. Med.*, 77, 144, 1971.

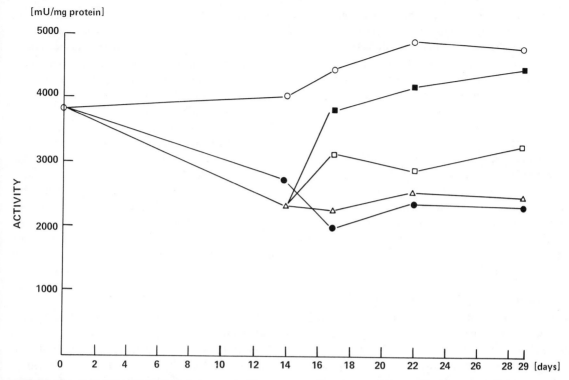

[mU/mg protein]

FIGURE 26.   Activities of latic dehydrogenase in liver of depleted, repleted, and control rats. (■) Zinc-repleted rats (12 ppm Zn); (●) pair-fed controls (96 ppm Zn); (○) control rats (96 ppm Zn), (□) zinc-repleted rats (4.5 ppm Zn); (△) zinc-deficient rats (1.2 ppm Zn). (From Kirchgessner, M. and Roth, H. P., *Arch. Tierernaehr.*, 25, 83, 1975. With permission.)

liver zinc concentration of zinc-deficient rats is about parallel to that of rapidly growing zinc-supplemented controls.[250]

Aldolase, NADH Diaphorase, and Pyridoxal Phosphokinase

While the zinc metalloenzyme aldolase exhibited no change in the rat,[240] its activity decreased in the kidneys and liver of zinc-deficient pigs.[262] Zinc-dependent NADH diaphorase also showed lower levels during zinc deficiency.[259] No change was found in the activity of the pyridoxal phosphokinase of the rat liver.[268]

It is customary to express the enzyme activities in terms of units per milligram of protein. However, Prasad and Oberleas[240] pointed out that impaired growth is one of the major effects of zinc deficiency; therefore, the protein content of the cells of certain body tissues may be altered. Thus, it is unsatisfactory to express enzyme activities per milligram of protein. Therefore, these authors suggested expressing enzyme activities in terms of DNA, which is a most stable cellular constituent.

On the other hand, evidence indicates that zinc deficiency also affects the DNA and RNA content of tissues. As previously stated, the activities of alkaline phosphatase and pancreatic carboxypeptidase A were reduced before growth retardation and inappetence became evident. Thus, it seems justified to assume that the protein content of the tissues was not yet affected. Furthermore, evidence for the conclusion that the response of these enzymes is specific for zinc is provided by the observation that the activities of only zinc-dependent enzymes are altered by zinc deficiency as compared to control levels of pair-fed animals. Activities of zinc-independent enzymes were not different between zinc-deficient and pair-fed animals. This has been demonstrated for sorbitol dehydrogenase and manganese-dependent isocitric dehydrogenase, glutamic-pyruvic transaminase, glutamic-oxalacetic transaminase of rats,[263] and isocitric dehydrogenase and iron-dependent succinic dehydrogenase of rats and pigs.[240,262]

The results presented in this review clearly

reveal that only specific zinc metalloenzymes change their activity and only in certain tissues. Therefore, the probability of detecting biochemical changes is highest in those tissues which sensitively respond to a lack of available zinc, for example testes, bones, pancreas, and intestinal mucosa.[262]

It is not to be expected that zinc-dependent enzymes are affected to the same extent in all tissues of a zinc-deficient animal. Differences in the sensitivity of enzymes are evidently the result of differences in both the zinc-ligand affinity of the various zinc metalloenzymes and in their turnover rates in the cells of the affected tissues.[262] Thus, it is to be expected that those zinc metalloenzymes which bind zinc with a very high affinity are still fully active even in extreme stages of zinc deficiency.

Swenerton, Shrader, and Hurley,[266] and Swenerton and Hurley[268] could not find reduced activities of the lactic, glutamic, or alcohol dehydrogenases in the liver of zinc-depleted rats, nor were there reduced activities of the malic and lactic dehydrogenases in the testes showing histological lesions. Therefore, they did not favor the hypothesis that reduced enzyme activities are responsible for the severe physiological and morphological changes observed in zinc-deficient animals. Even when symptoms of severe zinc deficiency are apparent, the complete lack of responsiveness of the activity of these enzymes may again be explained on the basis that the affinity of zinc to these enzymes is high, consequently the turnover rate in these tissues remains unaltered.

Reduced Enzyme Activities and Symptoms Due to Zinc Deficiency

Severe zinc deficiency is associated with reduced activities of a number of zinc-containing enzymes. Since tissues bind zinc with different affinity, a dietary zinc depletion may rapidly lead to a deficiency in labile zinc, especially in certain organs; therefore, it is associated with a corresponding loss in the activity of specific zinc metalloenzymes. An adequate supplementation with zinc rapidly overcomes this deficiency and raises the activity of these zinc metalloenzymes to normal levels. The extent to which the metalloenzymes lose their activity also depends on the functional role that zinc plays in maintaining the

enzyme structure. In some zinc-dependent enzymes (e.g., the alkaline phosphatase) zinc deficiency may induce structural changes which increase the chance for degradation. The consequence is an increased turnover rate and a lower activity of the enzymes in the tissues.[269]

Mills et al.,[270] Prasad et al.,[262] and Prasad and Oberleas[271] suggested that the rapidity with which biochemical changes arise in response to zinc depletion and then disappear upon repletion helps to identify the primary site of metabolic functions of zinc. In studies applying dietary zinc depletion, the early changes in enzyme activities, which are detectable before a general depletion becomes evident from tissue zinc levels, indicate that the primary role of zinc must be associated with a tissue component of an extremely high turnover or that zinc is essential at a site where it is freely exchangeable.[260]

Many metabolic processes are regulated by zinc metalloenzymes, which in turn depend on the tissue levels of zinc available in the control of their synthesis and activity.[260,261] Pancreatic carboxypeptidase A and alkaline phosphatase are enzymes that reduce their activity before food intake and growth are affected or even before lesions appear. However, a causal connection has not yet been demonstrated between the reduced activities and the deficiency symptoms. At least inferences can be drawn from the changes in the enzyme activities regarding the state of supply and the zinc requirement of man and animal.[272,273] Furthermore, according to model studies described by Kirchgessner et al.,[274] responsive zinc metalloenzymes can be used to determine the "metabolic efficiency" of absorbed zinc for synthetic processes in metabolism.

Total Protein in Zinc Deficiency

Several investigators[262,275,276] have reported that the total protein content of various tissues of zinc-deficient rats was lowered compared to that of zinc-supplemented, pair-fed animals. Somers and Underwood[80] demonstrated that zinc-deficient rat testes contain a higher level of nonprotein nitrogen. In lambs the retention of dietary amino acids was reduced, in comparison to that of pair-fed control animals, as indicated by an increased urinary excretion of nitrogen and sulphur. Biochemical changes in protein metabolism are also indicated by abnormalities in the protein pattern of plasma and serum.[277-279] Zinc deficiency is also associated with abnormalities in amino acid metabolism. On the basis of their data, Macapinlac et al.[275] and Somers and Underwood[80] suggested that zinc deficiency stimulates protein catabolism. However, other reports indicate that protein synthesis is impaired,[280,281] and more recent investigations show that zinc deficiency induces changes in nucleic acid metabolism which may limit protein biosynthesis, growth, and tissue regeneration.

Changes in Nucleic Acid Metabolism

It was shown some time ago that zinc deficiency may affect the ribonucleic acid (RNA) and deoxyribonucleic acid (DNA) metabolism of microorganisms and plants and thus be responsible for impaired protein synthesis.[282-285] This has recently been demonstrated to be true for animals also.

Tissue Levels of RNA and DNA

The biochemical changes brought about by zinc deficiency in animals may be so extensive that they are indicated by gross alterations in the cellular composition of certain tissues. Thus, testes and connective tissue of zinc-deficient rats were found to contain significantly less RNA and DNA

than those of pair-fed and ad libitum-fed control animals.[80,145,275] In rapidly regenerating connective tissue, RNA was affected more than DNA, as indicated by a significantly lower RNA/DNA ratio.[145] Similarly, lower RNA/DNA ratios were observed in the liver, kidneys, and pancreas of zinc-deficient young pigs compared to pair-fed control animals, while the DNA content per weight of tissue remained largely unaltered[262] (see Table 25). In the thymus of the young pig, however, the DNA content seems to be consistently depressed.[286]

Catabolism and Synthesis of RNA and DNA

Recent studies provide evidence that such changes in tissue levels of RNA and DNA of zinc-deficient animals may be the result of both an increased catabolism and an impaired biosynthesis of these polynucleotides. As shown in Table 22, increased activities of ribonuclease were found in several tissues of the zinc-deficient rat.[80,287] Zinc at a concentration of 10^{-4} M was previously shown to completely inhibit the activity of yeast ribonuclease.[288] Thus, an enhanced ribonuclease activity may explain the lower RNA levels and RNA/DNA ratios observed in certain zinc-deficient animal tissues (Table 22). Although, at least in part, it may also account for the impaired protein

TABLE 22

The Effect of Zinc Deficiency on the Activity of Enzymes Involved in Nucleic Acid Metabolism of Animals

Enzyme	Species	Tissue	Enzyme activity[a]	Ref.
Ribonuclease (acid)	Rat	Testis	+	80
Ribonuclease (acid)	Rat	Testis, kidney, bone	+	287
Ribonuclease (alkaline)	Rat	Testis, kidney, thymus	+	287
Deoxyribonuclease (acid and alkaline)	Rat	Testis, kidney, thymus, bone	0	287
RNA Polymerase	Rat	Liver nuclei	−	305
DNA Polymerase	Rabbit	Kidney cortex (cell cultures)	−	309
Thymidine Kinase	Rabbit	Kidney cortex (cell cultures)	−	309
	Rat	Connective tissue	−	239

[a]+, Increased; 0, unchanged; −, reduced.

Modified from Kirchgessner, M., Roth, H. P., and Weigand, E., in *Trace Elements in Human Health and Disease*, Vol. 1, Prasad, A. S., Ed., Academic Press, New York, 1976, 216. With permission.

synthesis and growth retardation (which are common manifestations of zinc deficiency), this response is not an early, sensitive effect.[287] Deoxyribonuclease activity assayed in testes, kidney, bone, and thymus of zinc-deficient rats was comparable to the pair-fed control animals.[287]

Plasma RNase activity was increased[120] in sickle cell disease patients (an example of conditioned deficiency of zinc). This observation suggests that measurement of plasma RNase activity may be a helpful diagnostic test for zinc deficiency in man.

Several investigators have studied the effects of zinc deficiency on the biosynthesis of RNA and DNA. There is sufficient evidence to implicate essential functions for zinc in both RNA and DNA synthesis, although some investigators failed to find differences in the in vivo incorporation of precursors into the polynucleotides of the testes and brain of zinc-deficient and restricted-fed zinc-supplemented rats.[275,289]

Livers from zinc-deficient rats incorporated less phosphorus-32 into the nucleotides of RNA and DNA than livers from pair-fed control animals.[290] Zinc injected intraperitoneally into partially hepatectomized rats stimulated the incorporation of labeled orotic acid into rapidly synthesized nuclear RNA.[291,292] Zinc was found to be necessary for RNA and DNA synthesis in monolayer animal cell cultures.[293,294]

There are numerous studies showing that zinc deficiency in animals impairs the incorporation of labeled thymidine into DNA.[145,165,239,291,296-303] This effect has been detected within a few days after the zinc-deficient diet was begun.[30,239,301] Williams and Chester[298,299] observed a progressive fall in the thymidine incorporation into DNA of liver, kidney, and spleen within 5 days after the zinc-deficient diet was fed to rats. Thus, dietary zinc deficiency may result in an immediate impairment of DNA biosynthesis. Prasad and Oberleas[239] provided evidence that decreased activity of thymidine kinase may be responsible for this early reduction in DNA synthesis and may ultimately relate to growth retardation (Tables 23, 24, and 25). As early as 6 days after the animals were placed on the dietary treatment, the activity of thymidine kinase (an enzyme essential for DNA synthesis) was reduced in rapidly regenerating, connective tissue of zinc-deficient rats, compared to pair-fed controls. These results have recently been confirmed by Dreosti and Hurley.[304] The activity of thymidine kinase in 12-day-old fetuses taken from females exposed to a dietary zinc deficiency during pregnancy was significantly lower than in ad libitum and restricted-fed controls.[304] Activity of the enzyme was not restored by in vitro addition of zinc, whereas addition of copper severely affected the enzyme activity adversely.[304]

As summarized in Table 22, the activities of other enzymes involved in polynucleotide synthesis are also affected by zinc deficiency. In liver nuclei from suckling rats nursed by dams on a zinc-deficient diet, the DNA-dependent RNA polymerase activity increased very little from birth to the 10th day of life and then started to fall, while in pups nursed by pair-fed dams the activity of this enzyme was not suppressed.[305]

TABLE 23

Thymidine Kinase Activity in Regenerating Tissue (nm TMP formed/hr/mg protein)

Days on diet	Deficient (A)	Pair fed (B)	Ad libitum (C)
6 day[a]	1.04 ± 0.14 (12)	3.57 ± 0.36 (13)	3.37 ± 0.36 (12)
13 day[b]	0.58 ± 0.02 (6)	2.40 ± 0.59 (5)	1.65 ± 0.12 (5)
17 day	None (5)	2.70 ± 0.6 (5)	2.68 ± 0.7 (5)

Note: ±, mean ± SE.

[a]6 day, deficient vs. either control $p < 0.001$.
[b]13 day, deficient vs. either control $p < 0.025$.

From Prasad, A. S. and Oberleas, D., *J. Lab. Clin. Med.,* 83, 634, 1974. With permission.

TABLE 24

[14]C-Thymidine Incorporation into DNA (6-day experiment DPM/mg DNA)

	Deficient	Pair fed	t[a]
Experiment I	$18.0 \pm 5.7 \times 10^3$ (4)	$136.2 \pm 15.2 \times 10^3$ (4)	7.24+
Experiment II	$14.4 \pm 2.9 \times 10^3$ (6)	$54.7 \pm 7.5 \times 10^3$ (5)	4.99+

Note: ±, mean SE.

[a]$p < 0.001$.

From Prasad, A. S. and Oberleas, D., *J. Lab. Clin. Med.,* 83, 634, 1974. With permission.

In liver nuclei of rats that had been maintained on normal diets, zinc was found to increase the RNA polymerase activity in vivo.[292] The DNA-dependent RNA polymerase from *Escherichia coli* was shown to be a zinc metalloenzyme.[306] According to studies by Slater, Mildvan, and Loeb[307] and Springgate, Mildvan, and Loeb,[308] zinc is also an essential constituent for the DNA polymerase of *E. coli,* sea urchins, and T^4-bacteriophages. Lieberman et al.[309] demonstrated that the activity of DNA polymerase of kidney cortex cells cultured from the rabbit did not increase when zinc was depleted by adding EDTA to the medium. The inhibition of DNA synthesis in animal cells affected by EDTA can be reversed by the addition of zinc.[294,309]

Polynucleotide Conformation

Although it appears that zinc has its primary effect on zinc-dependent enzymes that regulate the biosynthesis and catabolic rate of RNA and DNA, it cannot be ruled out that zinc is also associated more directly with the nucleic acids. Zinc may play a role in the maintenance of polynucleotide conformation.[310,311] Sandstead et al.[276] and Sandstead, Hollaway, and Baum[312] observed abnormal polysome profiles in the liver of zinc-deficient rats and mice. Acute administration of zinc appeared to stimulate polysome formation both in vivo and in vitro.[312] This finding is supported by the data of Fernandez-Madrid, Prasad, and Oberleas,[145] who noted a decrease in the polyribosome content of zinc-deficient connective tissue from rats and a concomitant increase in inactive monosomes. Using normal rats Weser, Hübner, and Jung[313] demonstrated an enhanced formation of high molecular weight and/or ribosomal RNA, which

was isolated 10 hr after the intraperitoneal injection of labeled zinc.

Zinc and Hormones

Since the discovery by Scott[314] that crystalline insulin contains considerable amounts of zinc (~0.5%), many studies have investigated the extent to which zinc nutrition of an animal influences the zinc content of the pancreas and its production, storage, and secretion of insulin. Inactive proinsulin is synthesized by the β-cells of the islets of Langerhans of the pancreas. By proteolytic cleavage of the connecting peptide, insulin is released to greatly influence carbohydrate, lipid, and protein metabolism.

Glucose Tolerance

One of the best known functions of insulin is to lower the blood glucose level. As early as 1937 Hove, Elvehjem, and Hart[315] published their first studies on the glucose tolerance of zinc-deficient rats. They noted only minor differences in the glucose tolerance curves after oral glucose doses between zinc-deficient and ad libitum-fed control rats. Hendricks and Mahoney[316] found no difference between zinc-deficient and zinc-supplemented rats in their inability to metabolize orally administered glucose. However, when glucose was injected intraperitoneally into rats that fasted overnight after a long period of dietary treatment as was done by Quarterman, Mills, and Humphries,[317] the glucose tolerance of zinc-deficient animals was depressed compared to that of pair-fed controls. This finding was confirmed by Boquist and Lernmark[318] (using Chinese hamsters), by Hendricks and Mahoney,[316] and by Huber and Gershoff[319,320] (using rats).

Figure 27 shows glucose tolerance curves

TABLE 25

Gain in Body Weight of Rats, Sponge Connective Tissue, and Concentration of DNA, RNA, Protein, and Zinc in Sponge Connective Tissue in 6-day Experiments

	Total gain in body weight (g)	SCT dry weight (mg)	DNA (μg) per mg T	RNA (μg) per mg T	Protein (mg) per mg T	Zn (μg) per mg T
Zn deficient	18 ± 1.6 (24)[a]	117.8 ± 10.6 (16)	6.2 ± 0.39 (12)	11.3 ± 0.39 (12)	0.54 ± 0.03 (12)	0.10 ± 0.008 (12)
Pair-fed controls	14 ± 1.2 (24)	153.3 ± 16.2 (15)	7.54 ± 0.37 (12)	13.4 ± 1.0 (12)	0.56 ± 0.02 (12)	0.14 ± 0.007 (12)
Ad-libitum-fed controls	53 ± 2.0 (24)	176.5 ± 17.7 (15)	6.6 ± 0.26 (12)	14.6 ± 1.5 (12)	0.52 ± 0.03 (12)	0.12 ± 0.19 (12)
p value						
Zn deficient vs. PF	0.05	NS	0.025	NS	NS	0.005
Zn deficient vs. ad libitum	0.001	0.01	NS	NS	NS	NS
PF vs. ad libitum	0.001	NS	NS	NS	NS	NS

Note: SCT, sponge connective tissue; T, tissue; ±, mean SE.

[a]Number in parentheses indicates number of rats.

From Prasad, A. S. and Oberleas, D., *J. Lab. Clin. Med.*, 83, 634, 1974. With permission.

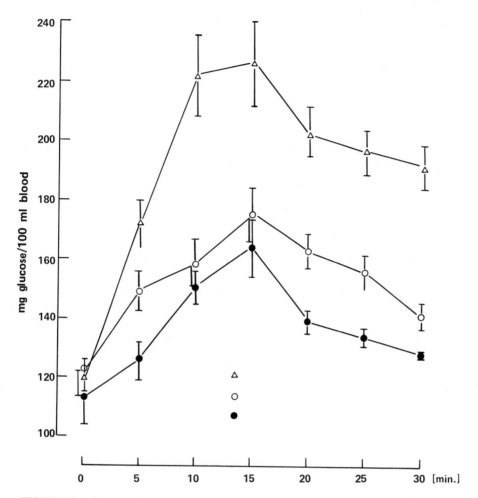

FIGURE 27. Glucose tolerance curves of zinc-deficient rats in comparison to ad libitum-fed and pair-fed control animals. The vertical bars represent the standard errors of the mean of six animals. (△) Zinc-deficient rats (2 ppm Zn); (○) ad libitum-fed rats (100 ppm Zn); (●) pair-fed rats (100 ppm Zn). (From Kirchgessner, M., Roth, H. P., and Weigand, E., in *Trace Elements in Human Health and Disease,* Vol. 1, Prasad, A. S., Ed., Academic Press, New York, 1976, 194. With permission.)

obtained by Roth, Schneider, and Kirchgessner[321] from zinc-deficient rats in comparison to pair-fed and ad libitum-fed control animals. In these studies rats which had been depleted by being fed a semisynthetic zinc-deficient casein diet (2 ppm zinc) for 34 days received an intramuscular injection of 80 mg of glucose per 100 mg of body weight after they had fasted for 12 hr. The zinc-depleted rats, which had the same initial plasma glucose concentration as the pair-fed and ad libitum-fed control animals, had a significantly lower glucose tolerance. Since the pair-fed animals exhibited a glucose tolerance that was even better than that of the ad libitum-fed controls, the lowered glucose tolerance of the zinc-deficient animals cannot be attributed to inanition.

In contrast to these findings is the report by Macapinlac, Pearson, and Darby,[242] who were unable to demonstrate that zinc deficiency affects the tolerance to intraperitoneally injected glucose, and the recent studies by Quarterman and Florence,[322] who used meal-eating and continuously eating pair-fed control rats. Quarterman and Florence suggested that the reduced glucose tolerance of zinc-deficient rats was merely the result of the different pattern in food intake, because zinc-deficient animals eat slowly and continuously throughout the day while their meal-fed pair mates consume their dietary allowance for the day in a rather short period of time. Thus, these authors consider the amount of food consumed on the day before the glucose-tolerance test is made to be the decisive factor. The conflicting results obtained for the glucose tolerance after oral dosing on the one hand, and after intraperitoneal, intravenous, or intramuscular injection on the other hand may be explained by a greater stimulation of insulin secretion by orally administered glucose.[323,324]

Insulin

The reasons for the poor glucose tolerance of zinc-deficient animals are not clear. Quarterman, Mills, and Humphries[217] demonstrated that zinc-deficient rats exhibit a reduced concentration of plasma insulin compared to pair-fed controls. They believe that the rate of insulin secretion in response to a glucose stimulation is reduced in zinc deficiency. Furthermore, the zinc-depleted animals were less sensitive to coma and convulsion when soluble, zinc-free insulin was injected intraperitoneally, although there were no differences in the blood glucose levels. Similarly, Huber and Gershoff[319,320] noted that the serum of zinc-deficient rats contained less immunoreactive insulin compared to that of ad libitum control animals; however, there was no difference when compared to the pair-fed controls. Total serum insulin-like activity, measured by in vitro adipose tissue assay, was significantly lower in the zinc-deficient groups than in the pair-fed and ad libitum control rats. They demonstrated in vitro that the pancreas from zinc-deficient rats incubated with glucose as stimu-

lant released less immunoreactive insulin as well as insulin-like activity.

In a further experiment, Quarterman and Florence[322] found no difference in the plasma insulin levels between zinc-deficient rats and their zinc-supplemented, continuously fed or meal-fed pair mates. Similarly, in four studies conducted by Roth and Kirchgessner,[325] the serum or plasma insulin levels of the zinc-deficient and pair-fed rats were significantly different only once though they were consistently lower compared to ad libitum-fed controls. Table 26 presents the blood glucose and plasma insulin levels of zinc-deficient and zinc-supplemented, pair-weight rats after a dietary treatment for 20 days and 15 min after a single or double intramuscular injection of 80 mg of glucose per 100 g of body weight. Again, the glucose tolerance in the zinc-deficient rats was significantly lower than in the pair-weight controls. The plasma insulin levels increased after the glucose injection, but were not different in zinc-depleted and control animals. Although there may be no difference in the insulin level, as determined by radioimmunoassays, it is possible that the physiological potency of the hormone is reduced by zinc deficiency. One must also distinguish between free active and bound inactive insulin in circulation. Zinc deficiency might affect these forms differently.

Although they observed a lowered glucose

TABLE 26

Blood Glucose and Plasma Insulin Levels of Zinc-deficient Rats and Pair-weight Control Animals

	Glucose level (mg/100 ml)		Insulin level (μU human insulin equiv./ml)	
	Zinc-deficient rats	Pair-weight controls	Zinc-deficient rats	Pair-weight controls
Initial levels	99.4 ±3.1[a]	98.6 ±3.2	5.7 ±0.4	4.7 ±0.3
15 min after one injection	165.9 ±4.4	124.8 ±9.9	13.0 ±1.6	10.7 ±1.9
15 min after two injections	196.9 ±7.9	159.1 ±7.9	13.0 ±2.1	9.6 ±2.3

[a]SE of the mean for 7 animals.

From Kirchgessner, M., Roth, H. P., and Weigand, E., in *Trace Elements in Human Health and Disease*, Vol. 1, Prasad, A. S., Ed., Academic Press, New York, 1976, 196. With permission.

tolerance, Boquist and Lernmark[318] did not find a reduced serum insulin concentration before or after the intravenous administration of glucose to zinc-deficient hamsters. Since a similar reduction in glucose tolerance was found after pancreatectomy[326] and after the administration of alloxan,[327] they believed that zinc deficiency causes a "prediabetic" condition. Furthermore, light and electron-microscopic studies showed that the β-cells of the pancreas exhibit a reduced granulation and thus possibly a reduced insulin content. Engelbart and Kief[328] found that acute stimulation of insulin secretion in rats also reduces the zinc content in the β-cells of the pancreas. Since it can be assumed that zinc participates in the synthesis and storage of insulin in the β-cells, it is plausible that the amount of insulin stored during zinc deficiency is lower. Coombs, Grant, and Frank[329] demonstrated (using equilibrium dialysis) that porcine proinsulin aggregates to soluble polymers by binding 5 M of Zn^{++} per mole while the porcine insulin binds only 1 M of Zn^{++} per mole and precipitates from solution. On the other hand, Hendricks and Mahoney[316] postulated that the reduced glucose tolerance of zinc-deficient animals is caused by an increased rate of insulin degradation. This could also explain the increased insulin resistence of zinc-depleted rats observed by Quarterman, Mills, and Humphries.[317]

In order to clarify the relationship between zinc deficiency and insulin, further research is needed with respect to the synthesis, function, and degradation of active insulin.

Growth and Sex Hormones

Homan et al.[330] demonstrated that the addition of zinc salts increases and prolongs the physiological potency of corticotropin preparations. In in vitro studies with human cell cultures, adrenal steroid hormones with glucocorticoid activity increased the uptake of Zn^{++}.[331] On the other hand, injections of gonadotropin and testosterone stimulated the growth of the male accessory sex organs of zinc-deficient rats, but did not prevent the tubular atrophy of the testes, which is considered to be typical of the zinc-deficient state.[332] No conclusions with respect to pituitary functions and zinc deficiency could be drawn in human studies.

Treating zinc-deficient rats with bovine growth hormone did not result in any improvement in weight gains.[242,261] Similarly, Ku[333] reported that growth hormone given to zinc-deficient pigs did not improve growth and food intake and had no influence on the serum zinc level, the serum alkaline phosphatase activity, and the parakeratotic lesions. The administration of bovine growth hormone to zinc-deficient nonhypophysectomized rats in the studies by Prasad et al.[216] also failed to enhance growth, while the growth rates greatly increased after zinc supplementation. The growth rates of hypophysectomized rats, however, responded to both hormone and zinc supplementation irrespective of the zinc status. Here the effects of the hormone and zinc were additive, but independent of each other. Gombe, Apgar, and Hansel[334] observed that the content of the luteinizing hormone (LH) in pooled pituitaries of zinc-deficient female rats was not different from that of pair-fed and ad libitum-fed control animals. However, the levels of LH and progesterone were reduced in the plasma of the zinc-deficient, and also in the restricted-fed animals in comparison to that of the ad libitum-fed controls. The lower plasma LH levels of zinc-deficient rats and their restricted-fed mates do not seem to be due to a lack of the LH-releasing factor, since its level was comparable in all three groups.

Recently the role of zinc in gonadal function was investigated in rats.[335] The increases in LH, follicle-stimulating hormone (FSH), and testosterone were assayed following intravenous administration of synthetic luteinizing hormone-releasing hormone (LH-RH) to zinc-deficient and restricted-fed control rats. Body weight gain, zinc content of testes, and their weights were significantly lower in the zinc-deficient rats as compared to the controls. The serum LH and FSH responses to LH-RH administration were higher in the zinc-deficient rats, but serum testosterone response was lower in comparison to the restricted-fed controls. These studies indicate a specific effect of zinc on testes and suggest that gonadal function in the zinc-deficient state is affected through some alteration of testicular steroidogenesis.

It is evident that there is a great need for further research on the role of zinc in hormone metabolism, especially on its function in the synthesis and secretion of various hormones.

Zinc in Collagen Metabolism

Several investigations have suggested that zinc

plays a fundamental role in protein biosynthesis activity. For instance, zinc was required for synthesis in *Rhizopus nigricans*[336] and *Euglena gracilis*.[283] Since collagen is the main fibrous protein of the connective tissue and is largely responsible for the development of tensile strength in the healing wound, biochemical studies have concentrated on the question of whether there is a specific effect of zinc deficiency on collagen synthesis, hydroxylation, conversion of procollagen to collagen, or some other aspect of its metabolism. Indeed, in a study of zinc deficiency in the rat, Fernandez-Madrid, Prasad, and Oberleas[145] found a significant reduction in total collagen in sponge connective tissue, as compared to pair-fed controls (Table 27). In the same study there was also a reduction in the total dry weight of the sponge connective tissue and the noncollagenous protein content in the zinc-deficient tissue as compared with the pair-fed controls. Moreover, the RNA/DNA ratio was significantly lower in zinc-deficient connective tissue, and in more severe deficient states, there was also depletion of polyribosomes and a significant reduction of RNA as compared with the connective tissue of pair-fed rats (Table 28 and Figures 28 and 29).

These data clearly indicate that the effect of zinc deficiency on collagen deposition was a generalized effect on protein synthesis and nucleic acid metabolism, rather than a specific effect on collagen synthesis. In fact, no differences were found with respect to level of hydroxylation, ultracentrifugation, chromatography in CM-cellulose columns, or disc gel electrophoresis between highly purified collagen from zinc-deficient animals and pair-fed rats. Other studies have also attempted to answer the same questions. The work of Hsu, Anthony, and Buchanan[337] revealed that zinc deficiency drastically reduced the incorporation of labeled glycine, proline, and lysine into rat skin. This study also showed that there were no marked changes in the uptake of these amino acids into liver, kidney, testes, or muscle protein. Since collagen is unusually rich in these amino acids, it was suggested that perhaps zinc could be more important in the metabolism of skin collagen than in the metabolism of other proteins. McClain et al.[164] have tried to further elucidate the role of zinc in collagen metabolism. They found a decrease in the salt-soluble fraction obtained from zinc-deficient animals which was thought to be due to a reduction in protein synthesis. In support of that conclusion, they found a reduction in the incorporation of labeled glycine into α_1 and α_2 chains of salt-soluble rat skin collagen. They did not find a significant reduction in the incorporation of labeled leucine into muscle polyribosomes, but there was an overall reduction of the polysome yield in the zinc-deficient animals. These data, in agreement with the abovementioned work, suggest that total collagen is reduced in the zinc-deficient state as a part of a generalized impairment in protein synthesis.

Elias and Chvapil[168] explored the question as to whether zinc supplementation affects the rate of collagen synthesis and the extent of collagen deposition in skin wounds in normal rats. They found that the administration of zinc did not affect the mechanical properties of the skin wound or total collagen content in the skin wound granulation tissue of normal rats. This lack of a stimulatory effect could be explained by assuming that normal rats were adequately supplied with the zinc required for connective tissue development; therefore, additional zinc supplementation was not necessary. However, they did find a stimulation of collagen synthesis and an increase in the extent of collagen deposition when the same experiment was done on a rat with a carbon tetrachloride-injured liver. Since experimental liver injury and various types of chronic illnesses are frequently accompanied by relative depletion of zinc,[34,91,141] these results are in agreement with the abovementioned reports of zinc stimulation of connective tissue development in the zinc-deficient state.[145,164]

Conflicting data have been presented by Waters et al.[338] These authors approached the question of whether zinc might accelerate wound healing by promoting an increased production of collagen in in vitro studies of fibroblasts in tissue culture media. In that study, diploid human fetal fibroblasts derived from skin and muscle were incubated with the addition of 10^{-4} to 10^{-8} M zinc sulfate to the experimental and control media. Newly established (low density), rapidly growing cultures were employed to evaluate the effect of zinc on fibroblast proliferation while confluent, stationary cultures were utilized to study the effect on collagen production. Basal levels of contaminating zinc in the experimental media were not reported. In order to determine the direct effect of zinc on the fibroblast, three parameters were evaluated:

TABLE 27

Final Body Weight of Rats, Connective Tissue Weight and Content of Zinc, RNA, and DNA of Sponge Connective Tissue (SCT).

Experiment	Final body weight (g)	SCT per rat (mg)	SCT weight/sponge weight	Total SCT zinc per rat (μg)	SCT zinc (μg/mg T)	Total SCT DNA per rat (mg)	SCT DNA (μg/mg T)	Total SCT RNA per rat (mg)	SCT RNA (μg/mg T)	RNA/DNA
Six day										
A. Zinc deficient	143 ± 3.0[a]	180 ± 19	0.9 ± 0.13	8.2 ± 1.8	0.056 ± 0.009	2.533 ± 0.390	14.2 ± 1.95	3.833 ± 0.699	20.1 ± 2.7	1.4 ± 0.07
B. PF	162 ± 2.0	263 ± 19	1.2 ± 0.16	18.5 ± 2.3	0.072 ± 0.009	2.681 ± 0.318	10.2 ± 1.08	5.569 ± 0.614	21.1 ± 1.7	2.1 ± 0.26
P value										
A vs. B	<0.001	<0.025	NS	<0.025	NS	NS	NS	NS	NS	<0.05
Ten day										
A. Zinc deficient	110 ± 2	260.1 ± 10.7	0.5 ± 0.0489	34.5 ± 8.7	0.196 ± 0.04	2.34 ± 0.27	8.9 ± 0.89	1.93 ± 0.16	7.4 ± 0.62	0.87 ± 0.12
B. PF	132 ± 5	479.1 ± 29.1	0.7 ± 0.051	57.3 ± 2.8	0.141 ± 0.02	4.62 ± 0.46	9.6 ± 0.79	6.33 ± 0.43	13.4 ± 0.54	1.42 ± 0.14
C. Ad lib	321 ± 7	601.3 ± 42.8	0.8 ± 0.032	92.8 ± 19.4	0.132 ± 0.03	6.30 ± 0.42	10.6 ± 0.80	9.24 ± 0.78	15.4 ± 0.55	1.47 ± 0.08
p value										
A vs. B	<0.005	<0.001	<0.025	<0.05	NS	<0.005	NS	<0.001	<0.001	<0.025
A vs. C	<0.001	<0.001	<0.001	<0.05	NS	<0.001	NS	<0.001	<0.001	<0.005
B vs. C	<0.001	<0.5	NS	NS	NS	<0.025	NS	<0.01	<0.01	NS

Note: PF, pair-fed control animals; ad lib, ad libitum-fed control animals; SCT, sponge connective tissue; T, tissue; NS, not significant; ±, mean SE.

[a] Eight animals per observations.

From Fernandez-Madrid, F., Prasad, A. S., and Oberleas, D., *J. Lab. Clin. Med.*, 82, 951, 1973. With permission.

TABLE 28

Protein and Collagen Content of Sponge Connective Tissue (SCT)

Experiment	Total SCT protein per rat (mg)	SCT protein (mg/mg T)	Total SCT collagen per rat (mg)	Total SCT collagen (mg/mg T)	Collagen/ protein	Soluble collagen (% total SCT collagen)			Insoluble collagen, % total SCT collagen
						0.15 M NaCl	1.0 M NaCl	0.5 M acetic acid	
Six day									
A. Zinc deficient	98.31 ± 14.62[a]	0.54 ± 0.03	10.15 ± 1.62	0.056 ± 0.005	0.104 ± 0.007	1.44 ± 0.51	0.96 ± 0.05	41.3 ± 1.5	56.0 ± 1.27
B. PF	195.1 ± 18.29	0.14 ± 0.04	21.53 ± 1.62	0.086 ± 0.008	0.114 ± 0.015	1.18 ± 0.11	0.63 ± 0.09	32.4 ± 1.5	74.6 ± 1.39
p value									
A vs. B	<0.01	<0.01	<0.001	<0.01	NS	NS	<0.025	<0.001	<0.001
Ten day									
A. Zinc deficient	188.6 ± 12.3	0.73 ± 0.04	27.6 ± 2.78	0.106 ± 0.008	0.148 ± 0.015	1.65 ± 0.29	1.19 ± 0.18	3.3 ± 0.59	93.9 ± 0.59
B. PF	337.9 ± 39.4	0.70 ± 0.05	50.1 ± 5.73	0.104 ± 0.009	0.151 ± 0.017	1.89 ± 0.07	0.99 ± 0.09	2.67 ± 0.50	94.4 ± 0.54
C. Ad lib	425.2 ± 29.3	0.71 ± 0.03	43.9 ± 3.04	0.073 ± 0.003	0.104 ± 0.007	3.56 ± 0.23	3.14 ± 0.19	9.4 ± 0.84	83.8 ± 0.91
p value									
A vs. B	<0.025	NS	<0.01	NS	NS	NS	NS	NS	NS
A vs. C	<0.001	NS	<0.005	<0.01	<0.05	<0.001	<0.001	<0.001	<0.001
B vs. C	NS	NS	NS	<0.01	<0.05	<0.001	<0.001	<0.001	<0.001

Note: PF, pair-fed control animals; ad lib, ad libitum-fed control animals; SCT, sponge connective tissue; T, tissue; NS, not significant; ±, mean SE.

[a] Eight animals per observations.

From Fernandez-Madrid, F., Prasad, A. S., and Oberleas, D., *J. Lab. Clin. Med.*, 82, 951, 1973. With permission.

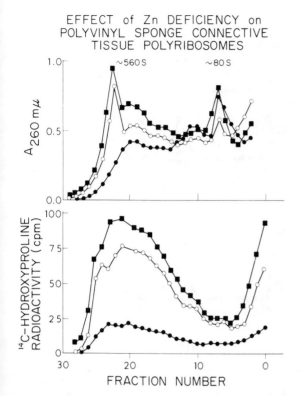

EFFECT of Zn DEFICIENCY on
POLYVINYL SPONGE CONNECTIVE
TISSUE POLYRIBOSOMES

FIGURE 28. Zone centrifugation through 15 to 60% linear sucrose gradients of polyvinyl sponge connective tissue polyribosomes. The capsules from polyvinyl sponges removed 10 days after implantation from zinc-deficient, pair-fed, and ad libitum-fed rats were incubated in vitro with ^{14}C-proline for 30 min. The ribosomal suspensions were centrifuged for 3 hr at 20,000 rpm in the SW 25.1 rotor of the Spinco® Model L ultracentrifuge. Absorbancy was monitored at 260 nm and total and ^{14}C-hydroxyproline radioactivity were determined in the individual fractions. (■) Ad libitum-fed control animals; (○) pair-fed control animals; (●) Zn-deficient animals. (From Fernandez-Madrid, F., Prasad, A. S., and Oberleas, D., *J. Lab. Clin. Med.*, 82, 951, 1973. With permission.)

1. Cellular proliferation, measured by changes in total culture protein

2. Rate of collagen biosynthesis, measured by the rate of conversion of radioactively labeled proline into hydroxyproline

3. Changes in culture collagen content, measured by total accumulation of hydroxyproline

The results showed that addition of zinc sulfate to low-density cultures in concentrations of 10^{-4} to 10^{-8} *M* produced concentration-dependent cytotoxicity with inhibition of collagen biosynthesis, as well as cellular proliferation. Similar addition of zinc sulfate at 10^{-4} *M* concentration to confluent

(high-density) cultures apparently resulted in significant inhibition of collagen biosynthesis. The authors interpreted these data to indicate that zinc does not have a direct acceleratory effect either on cellular proliferation or collagen biosynthesis in human fibroblasts in vitro. This interpretation of the data is not in agreement with several other studies suggesting the direct participation of zinc in the process of wound healing.[145,164,168] Unfortunately, the basal levels of zinc in the fibroblast cultures were not reported;[338] since zinc is a notorious contaminant, it is possible that their experimental culture media already had an adequate supply of zinc. Cytotoxicity was probably seen, rather than lack of a stimulatory effect. The possibility that the reduction in the extent of collagen deposition found in the zinc-deficient state may be related in part to an increased degradation has not been studied and, therefore, cannot be ruled out.

McClain et al.[164] have presented data which were interpreted as indicating a fundamental role for zinc in the process of cross-linking of collagen. Purified salt-soluble collagen from zinc-supplemented and zinc-deficient animals was subjected to disc gel electrophoresis. It was reported that the β-components from zinc-deficient animals were increased in comparison to those from zinc-supplemented animals. However, observation of the densitometric tracings of the disc gel patterns showed a minimal reduction of α_1 and practically identical content of α_2 chains in the zinc-deficient collagen as compared with that of zinc-supplemented animals. Also, the aldehyde content of the salt-soluble collagen from zinc-deficient animals was reported to be almost twice as high as that from the zinc-supplemented animals. These results were interpreted as indicating a greater degree of intramolecular cross-linking in the salt-soluble fraction of zinc-deficient skin collagen. It was speculated that the antagonistic relationship between zinc and copper could be the basis of the postulated cross-linking abnormality. Since copper is known to be required for lysyl oxidase (the enzyme which oxidatively deaminates the precursor of the collagen intramolecular cross-link), an increased activity of this enzyme in the zinc-deficient state might enhance formation of a covalent intramolecular cross-link.[164] Indeed, antagonism between copper and zinc has been reported.[339-341] In general, zinc administration decreased the level of copper in the

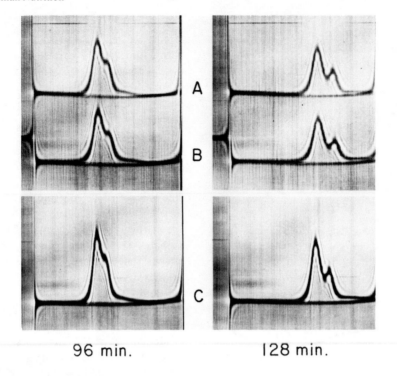

96 min.   128 min.

FIGURE 29.   Sedimentation patterns of denatured acid-extracted polyvinyl
sponge connective tissue collagen; A, ad libitum control animals; B, Zn-deficient
animals; C, pair-fed control animals. The samples were dissolved in 0.15 *M* sodium
formate, pH 3.75, and incubated for 15 min at 40°C. The undissolved material
was removed prior to ultracentrifugation by centrifuging at 2000 rpm for 15 min.
Ultracentrifugation was done with a Spinco Model E, equipped with an electronic
speed control, and An-D rotor with double-sector cells, with plain and positive
wedge windows. Centrifugation was carried out for 128 min at 60,000 rpm at
39°C. Percentages of α- and β-chains were as follows: A, 68.3 and 23.8; B, 63.4
and 26.0; C, 62.6 and 24.6. Higher aggregates in three samples were calculated as
follows: A, 7.9%, B, 10.6%, C, 12.9%. (From Fernandez-Madrid, F., Prasad, A. S.,
and Oberleas, D., *J. Lab. Clin. Med.,* 82, 951, 1973. With permission.)

serum and tissues. Also, the addition of zinc in
vitro has been shown to interfere with the activity
of lysyl oxidase.[342] It is also known that the
tissue concentration of copper increases in the
zinc-deficient state.[259] However, there is no
information in the literature about the activity of
copper-dependent enzymes in the zinc-deficient
state. In addition, accurate quantification of intramolecular cross-links cannot reliably be
accomplished by the determination of the proportion of α to β chains or the aldehyde content of
purified collagen. These data are of doubtful
significance and cannot be interpreted as an
expression of a cross-linking abnormality.

McClain et al.[164] also reported that the acid-
soluble collagen pool from zinc-deficient animals
was increased by nearly 20%, although the α/β
ratios did not vary markedly between the supple-

mented and deficient animals. The aldehyde
content of the acid-soluble collagen was not
reported; however, that of the insoluble skin
collagen was found to be 47% lower in the
zinc-deficient than in the zinc-supplemented
animals. The authors interpreted these data as
evidence of an inhibition of the intermolecular
cross-linking mechanism induced by the deficiency
of zinc. They suggested from their studies that
zinc has a fundamental role in the process of
cross-linking of collagen. Since solubility studies
are crude indicators of cross-linking abnormalities
and no differences were found in the acid-soluble
collagen by ultracentrifugation, disc gel electrophoresis, or CM-cellulose chromatography,[145,164]
it was concluded that these data do not contribute
evidence for a cross-linking defect.

Early studies of developing sponge connective

tissue have shown that a spurt of DNA synthesis precedes the increase in the deposition of collagen.[343] Indeed, at the time in which fibroblast proliferation as expressed by DNA synthesis is very active, there is very little accumulation of fibrous collagen in the sponge.[344] It seems that this initial period of cell division preceding collagen deposition is a common denominator of developing connective tissue in most circumstances.[343] Therefore, it is clear that any interference with the synthesis of DNA and fibroblast proliferation will profoundly influence the overall deposition of fibrous collagen in the developing connective tissue. Many studies have suggested that DNA synthesis may be impaired in zinc deficiency.[145,165,239,291,296-303] It remains to be established whether one enzyme is finally singled out as responsible for the inhibition of DNA synthesis or whether this effect is due to a constellation of enzymatic defects dependent on zinc. However, it is obvious that fibroblast proliferation is impaired and this defect is a major contributing factor to the abnormalities in wound healing found in the zinc-deficient state.

Zinc and Cystine Metabolism

A specific role of zinc in cystine metabolism has been observed.[345] The increased expired $^{14}CO_2$ after injection of $[1-^{14}C]$-DL-cystine in zinc-deficient rats may be linked to a defect in the utilization of this amino acid for protein synthesis. Such a possibility finds support from the existence of a reduction of 30% in the incorporation of $[1-^{14}C]$-DL-cystine into the liver and kidney proteins of zinc-deficient rats.[345]

Zinc deficiency is also associated with enhanced urinary total ^{35}S, which is comprised of inorganic $[^{35}S]$-sulfate ethereal $[^{35}S]$-sulfate and neutral $[^{35}S]$-sulfate after $[^{35}S]$-cystine injection. This finding is in harmony with a recent report by Somers and Underwood,[346] indicating that an increased urinary excretion of sulfur occurred in zinc-deficient ram lambs. The mechanism for these observations is not clearly understood at the present time. A possible explanation is that zinc may be involved in the utilization of inorganic sulfate.[114] The normal activity of liver ATP sulfurylase in zinc-deficient rats fails to support this view. Nevertheless, the effect of zinc deficiency on the reactions leading to incorporation of active sulfate into organic molecules requires further experimentation.

It has been shown that protein is constantly being degraded to amino acids, and protein is synthesized from amino acids at the same time. $[^{35}S]$-L-cystine is no exception. Thus, the increase of urinary excretion of radiosulfur and inorganic sulfate-^{35}S in zinc-deficient rats after the first 24-hr period following cystine-^{35}S injection suggests an increase of protein-^{35}S catabolism.[345] It is also possible that nonprotein compounds such as glutathione, in which cysteine is one of the precursors, are broken down at a rapid rate in zinc-deficient rats.[345] Further studies are needed to define the role of zinc in protein synthesis and catabolism before one can interpret the significance of metabolic changes in amino acids due to deficiency of zinc.

Effect of Zinc on Cell Membrane

The first indication that zinc modifies the cell plasma membrane was derived from the work of cytologists who used zinc to isolate intact cell membranes.[347] Although no direct evidence was presented on the mechanism of the zinc effect, it was assumed that zinc interacts with thiol groups at the membrane under the formation of stable mercaptides. The same mechanism was implicated in the stabilizing effect of zinc on neurotubules from the rat brain.[348] In this case, only the zinc and cadmium helped to isolate intact tubules, and zinc treatment produced tubules similar in structure to the ultrastructure of well-preserved neurotubules.

Peritoneal macrophages, isolated from mice that were injected intraperitoneally (i.p.) with thioglycolate and treated at the same time i.p. or intramuscularly with zinc (0.25 mg or 0.5 mg $ZnCl_2$ per mouse per day) showed a significantly higher viability index than macrophages from control animals.[349] When zinc-treated macrophages were exposed to silica particles (1 μm for 30 min), the cytotoxic effect of silica was significantly diminished. The reason for the decreased cytotoxicity of zinc-treated macrophages is that these cells display much lower phagocytosis of *Staphylococcus albus* than control macrophages.[357] Similar results have been obtained using young rats and guinea pigs fed (for a prescribed time) diets with various zinc contents amounting to 0.5, 40, and 2000 ppm.[357] A striking difference was noted in the migration capacity of macrophages isolated from animals fed the three different zinc diets (Figure 30). Macro-

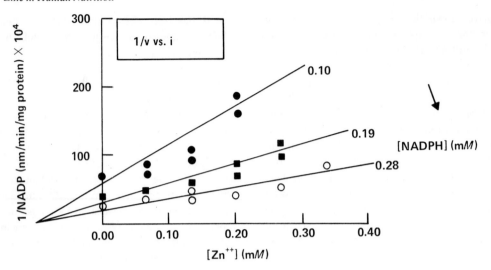

FIGURE 30.   Effect of zinc on the activity of NADPH oxidase in alveolar macrophages from dog lungs. The data show the inhibiting effects of Zn^{++} on the production of NADP as a function of NADPH concentration. The effect appears to be a competitive inhibition of zinc with respect to NADPH, (From Chvapil, M., Zukoski, C. F., Hattler, B. G., Stankova, L., Montgomery, D., Carlson, E. C., and Ludwig, J. C., in *Trace Elements in Human Health and Disease,* Vol. 1, Prasad, A. S., Ed., Academic Press, New York, 1976, 274. With permission.)

phages from guinea pigs fed a zinc-deficient diet (0.5 ppm zinc) displayed maximum mobility. They migrated statistically faster than those from animals fed 40- or 2000-ppm zinc diets. In several experiments, macrophages from animals fed a high-zinc diet did not migrate at all. Still, these cells showed significantly higher viability than cells from either one of the two lower zinc groups. Based on these results, it was concluded that a high-zinc diet functionally immobilized macrophages.[357]

Light microscopy of macrophages from control guinea pigs fed 40 ppm zinc showed most of the cells with extrusions, long-reaching pseudopodia, or ruffling plasma membrane.[357] Unlike these cells, macrophages from high-zinc-treated animals were round shaped, only a few showed minimal ruffling, and long extrusions or pseudopodia were completely absent. Scanning microscopy provided further evidence that a high-zinc diet inactivated the mobility of plasma membranes of peritoneal macrophages.

It has been hypothesized that zinc may paralyze the macrophage by interfering either with some enzymes regulating the activity of plasma membranes or with some functional groups of membrane constituents.[357] Thus, zinc appears to change the fluidity of this system.

Some evidence has been presented to show that

zinc inhibits the activity of various forms of ATPase in alveolar macrophages[350,351] or in brain microsomes.[352,353] Mustafa, Cross, and Hardie[350] also showed that the ATPase system is membrane bound and located predominantly in pulmonary alveolar, macrophage and cellular surface membranes. These results were reproduced and it was found that 0.5 mM zinc completely inhibited the activity of ATPase in activated macrophages. If an active Na^+-K^+-stimulated, Mg^{++}-dependent ATPase system is essential for cell membrane functions such as phagocytosis and pinocytosis and eventual active ionic-transport processes,[350] then the inhibitory effect of zinc on these functions seems to be justified. However, there are other possible mechanisms involved. The activity of macrophages has been linked to the activity of NADPH oxidase[354] as part of the phagocytosis process. An excellent review by Sbarra et al.[355] indicates that the enzymatic mechanism responsible for the stimulated metabolic activities, that accompany phagocytosis, appears to be directly involved with the oxidation of NADPH by its oxidase. Increased NADPH oxidase activity supplies pyridine nucleotides as well as forms of H_2O_2 needed for the bacteriocidal activity of the phagocytes.

In a detailed study of the zinc effect on NADPH oxidase activity in liver microsomes and in the alveolar macrophages, it was found that zinc competitively inhibits NADPH oxidase. In the system of liver microsomes, approximately 10 μM zinc inhibits the NADPH oxidase by 50%; in the whole homogenate of alveolar macrophages, the concentration of zinc required for 50% enzyme inhibition was higher (amounting to 120 μM). Inhibition of NADPH oxidation by low concentrations of zinc indicates that zinc may be an important regulator of NADPH oxidase activity in vivo.

Zinc as an Integral Part of Plasma Membrane of Macrophages

It is becoming evident that small quantities of zinc are present in various cell membranes. Most of the membrane-bound zinc is linked to a distinct macromolecule constituent lipoprotein fraction.[356,357] In this fraction, zinc is bound to the lipid moiety, whereas in the whole membrane the distribution of zinc between the protein and lipid phase is approximately 45 to 55%. However, the most amazing finding is that inducing lipoperoxidative damage to polyunsaturated fatty acid components of the erythrocyte membrane by ultraviolet irradiation or by introducing free radicals into the system does not cause the release of zinc from the ghost pellet into the medium, although some proteins and lipids are released into the incubation medium.

Chvapil et al.[357] studied the effect of zinc and other metals on aggregation of platelets and release of H^3-serotonin, activated either by collagen or epinephrine. They observed the following:

1. Aggregation of platelets in dog and human platelet-rich plasma (PRP), induced by either collagen or epinephrine, is inhibited by zinc within a narrow range of concentrations, between 5 and 15 μM zinc.

2. Release of H^3-serotonin from dog platelets is also significantly inhibited by zinc within the same concentration range.

3. The inhibitory effect of zinc on the release reaction depends on the presence of plasma or some proteins, mainly fibrinogen. Even at the lowest concentration, zinc added to a suspension of washed platelets in Hank's medium enhanced the release of H^3-serotonin, and the addition of fibrinogen restored the inhibitory effect of zinc.

4. Some other metals such as Cd and Mn were also inhibitory.

In a recent publication, Sacchetti et al.[358] studied the effect of manganese ions which were shown to displace Ca^{++}, thus modifying functions of the platelet membrane. It remains to be established whether or not the role of zinc is similar to that of manganese with respect to its effect on platelet membrane.

Lymphocytes and Zinc

This topic has been studied by several authors,[359-362] and their results can be summarized as follows: Administration of zinc in incubation media stimulates DNA synthesis of lymphocytes within 6 to 7 days; at this time, approximately 10% of DNA synthesis in cultured lymphocytes was maximal within a narrow range of zinc concentrations, varying between 0.1 and 0.2 mM. Among several cations tested, only zinc and mercury were stimulatory. Zinc must be present in the media for the entire culture period in order to produce maximal stimulation of H^3-thymidine incorporation in DNA lymphocytes.

Prasad and Oberleas[240] have reported that thymus may be a zinc-sensitive tissue. Pneumotropic viral infections and secondary bacterial infections (e.g., *Salmonella* species, *Pasteurella*, and *Pyrogenic* bacteria) have been noted to lead to rapid death in many cases of hereditary zinc-deficient Dutch Friesian cattle. These infections are probably related to faulty development of the immune system.[363] Absolute lymphopenia has been noted in patients with cirrhosis of the liver (an example of a conditioned zinc-deficient state) and in some species of animals made zinc deficient.[72] Frost et al.[363a] have reported that patients with sickle cell disease (also an example of a conditioned zinc-deficient state) have a considerable suppression in peripheral T-cell numbers and an increased number of null cells. In addition, these patients, while showing normal mitogen responsiveness in vitro, demonstrate an impaired cellular immune response in vivo as measured by skin tests.

Recent studies in rats indicate that zinc deficiency causes a selective suppression of lymphoid organ weight and abnormalities in the immune response to sheep erythrocytes.[363b] The overall response to this antigen is markedly depressed and delayed. Further studies must be conducted in order to define more precisely the role of zinc in immune responses.

Effects on Red Cells

One in vitro model with which to test anti-sickling agents is filterability. A 3.0-μm Nucleopore® filter challenges the deformability of red cells in somewhat the same manner as does a tissue capillary. In a study of filterability, incubation of whole blood at 4° for 2 hr with appropriate concentrations of zinc has been carried out with controls containing no zinc similarly handled.[364] After incubation, samples were diluted approximately 1:100 in buffered saline at pH 7.4 and equilibriated with appropriate gas mixtures; filterability was carried out at 37°C with a micron Nucleopore filter and 10-cm negative water pressure. The measurement of filterability was the length of time for 2.0 ml of the cell suspension to pass through the filter. It was observed that zinc in a concentration as low a 0.3 mM, but not lower, improved filterability of sickle cells at intermediate (15 to 30 mmHg) oxygen tensions.[364] Assay of zinc in red cells revealed that the amount of additional zinc incorporated after incubation with 0.3 mM zinc under these conditions was equal to a zinc/hemoglobin ratio of less than 0.01. Improvement in filterability at such low concentration suggested that the effect was not directly on hemoglobin and, as such, the effects of zinc on red cell membrane were investigated.[365]

It was observed that zinc markedly decreased the amount of hemoglobin retention by red cell membrane.[365] Single-stage red cell ghosts, prepared by the hypotonic lysis technique of Hoffman[366] were studied with agents incorporated into the membranes during a single 20-min hypotonic exposure. After resealing and repeated washing, the hemoglobin levels of the ghost preparations were measured. Zinc causes a six- to sevenfold reduction in the amount of hemoglobin normally retained by these ghosts.[365] It has previously been shown that calcium increases the amount of hemoglobin retained by single-stage ghosts.[367,368] Zinc partially blocks the hemoglobin-retaining effect of calcium on single-stage ghosts.[365] The effects of the presence of zinc and/or calcium on hemoglobin retention by ghosts prepared by the single stage method are summarized in Table 29.

Calcium accumulation in red cells greatly decreased their deformability, and recent studies by Eaton et al.[369] have implicated calcium incorporation as a pathogenic event in the formation of irreversibly sickled cells. The effect of zinc on calcium incorporation into intact sickled cells has been studied, and it was found that the amount of calcium incorporation is decreased about twofold in the presence of zinc. A better understanding of this observation may arise from the studies on the effect of zinc on ^{45}Ca incorporation into two types of red cell ghosts. The first type of ghosts was prepared by the single-stage technique, as described above, in the presence of 11 mM $CaCl_2$ solution labeled with ^{45}Ca. In three separate experiments, the number of calcium atoms incorporated per ghost was consistently decreased by zinc.[364] However, the differences in hemoglobin retention in the various preparations could have affected relative calcium retention. Therefore, ghosts were also prepared by reducing the hemoglobin content to a minimum in all preparations by multiple washing of unsealed ghosts. Incorpora-

TABLE 29

Hemoglobin Retention by Ghost Cells

Ghosts prepared in the presence of	Hemoglobin (g %) in final ghosts[a]	Retention of hemoglobin (%)[a]
Saline (control)	1.86	7.4
1.5 mM ZnSO$_4$	0.25	1.0
1.0 mM CaCl$_2$	6.39	24.6
1.5 mM ZnSO$_4$ + 1.0 mM CaCl$_2$	2.50	10.7

[a]Corrected for change in mean ghost cell volume. Average of four experiments.

Modified from Dash, S., Brewer, G. J., and Oelshlegel, F. J., Jr., *Nature*, 250, 251, 1974. With permission.

tion of ^{45}Ca per milligram dry weight of such hemoglobin-free ghosts was also consistently decreased in the presence of zinc.[364]

The effect of zinc on calcium binding and/or hemoglobin retention of red cell membranes may be involved in the beneficial effect of zinc on filterability of sickle cells. It is becoming clearer that the process of formation of irreversibly sickled cells involves the cell membrane. Calcium and/or hemoglobin binding may promote the formation of irreversibly sickled cells, thus hindering the filterability of such cells. Zinc may act favorably on the filterability of sickled cells by blocking the proposed calcium and/or hemoglobin binding to the membrane.

Zinc as a Viral Inhibitor

Trace elements serve a well-known role in biological systems, often appearing as cofactors in enzymes, membranes, and other macromolecular structures. Some of these elements may have a control function, since they readily inhibit certain enzymatic reactions.[370] This implies that trace elements could play a significant role in various features of virus replication in susceptible host cells. However, little attention has been paid to the effects of trace elements in the replication process of mammalian viruses.

The rhinoviruses (common cold viruses) are a large group of very small, simple human parasites. These infectious viruses are composed of a single strand of RNA enclosed in a protein coat or capsid.[371] They are able to form visible sites of infection on the host cells unless an inhibitor is present. Eight out of nine human rhinoviruses were studied and found to be susceptible to 10^{-4} M zinc ion.[371,372] One rhinovirus (type 5) as well as two serotypes of polio virus were resistant.[372]

There are two probable ways by which zinc may block cleavage of rhinovirus proteins.[371,372] One is by activation of one or more proteases; the other is by binding to and altering the substrate so that it cannot be cleaved. The latter model is preferred for the following reasons:

1. Zinc almost immediately blocks virus production, suggesting that one of the components of the virion is affected directly by zinc.
2. Zinc interacts directly with rhinovirus capsids.
3. Whereas sufficient amounts of purified

virus will produce crystals, amorphous precipitates form in the presence of a small amount of zinc.
4. The results of cleavage inhibition indicate that the sensitive proteolytic reactions invariably involve precursors containing capsid protein sequences.[373]

These data suggest that zinc ions bind to rhinovirus capsid polypeptides, prevent their successful combination with viral RNA, and block their nascent cleavage. It remains to be elucidated whether or not these observations have any significance with respect to viral infections in man.

Zinc and Metallothionein

Metallothionein is a low-molecular-weight protein.[383,384] rich in titrable sulfhydryl groups. It binds both cadmium and zinc. The molecular weight of metallothionein from equine renal cortex is approximately 10,000 and contains 5.9% cadmium, 2.2% zinc, 0.2% iron, and 0.1% copper. The sulfur content of the metal-free protein is 9.3%, and 95% of all sulfur is present in the form of sulfhydryl groups of cysteine. Human renal cortex metallothionein protein appears to be quite similar to the equine protein with respect to the physicochemical properties.

Recent studies indicate that the synthesis of metallothionein is controlled at the transcriptional level and is regulated by changes in the pool of translatable thionein in RNA by body zinc status.[385,386] It has been suggested that, in the liver, metallothionein synthesis functions in uptake and storage of zinc in hepatocytes, whereas in the intestinal mucosal cells, this protein competes with the normal ligand involved in zinc absorption and thus may regulate the amount of available zinc for transfer to the plasma. Further research is required to establish the functional roles of metallothionein.

TOXICITY

Three types of toxic reactions to zinc have been reported in man. First, the "metal fume fever," characterized by pulmonary manifestations, fever, chills, and gastroenteritis, has been observed to occur in industrial workers who are exposed to fumes.[374] In the second type, toxicity was observed in a 16-year-old male who ingested 12 g of zinc sulfate over a period of two days. This was characterized by drowsiness, lethargy, and increased serum lipase and amylase levels.[375] The third type of acute zinc toxicity has been observed in a patient with renal failure following hemodialysis. (The water for hemodialysis was stored in a galvanized tank.) The patient suffered from nausea, vomiting, fever, and severe anemia.[376]

Zinc, relatively speaking, is nontoxic in comparison with other trace elements. Many of the toxic effects attributed to zinc in the past may actually be due to other contaminating elements such as lead, cadmium, or arsenic.[377] Zinc is noncumulative, and the proportion absorbed is thought to be inversely related to the amount ingested.[378] Vomiting, a protective phenomenon, occurs after ingestion of large quantities of zinc.[379] In fact, 2 g of zinc sulfate has been recommended as an emetic.[379]

The symptoms of zinc toxicity in humans include dehydration, electrolyte imbalance, abdominal pain, nausea, vomiting lethargy, dizziness, and muscular incoordination. Acute renal failure caused by zinc chloride poisoning was reported by Csata, Gallays, and Toth.[380] The symptoms occurred within hours after large quantities of zinc were ingested. Death is reported to have occurred after ingestion of 45 g of zinc sulfate.[381] This dose would be considered very massive, in view of the fact that the daily requirement of zinc for man is considered to be in the range of 15 to 30 mg/day.

In a limited trial, in patients with sickle cell disease, 660 mg of zinc sulfate was administered orally for nearly 1 year without adverse effects. On a longer term basis, the side effects of zinc administration for therapeutic purposes remain to be elucidated. Physicians must remain cautious in the use of zinc salts for prolonged periods.

In rats, ingestion of 0.5 to 1.0% of zinc results in reduced growth, anemia, poor reproduction, and decreased activity of liver catalase and cytochrome oxidase.[382] The latter are reversed by copper administration, thus indicating that excessive intake of zinc may cause copper deficiency.

In view of long-term clinical usage of zinc in therapeutic dosages for wound healing and sickle cell disease in man, one must remain alert for possible toxic effects. It is well known that zinc and copper compete with each other for similar protein-binding sites, and it is conceivable that copper deficiency may be induced in subjects receiving high amounts of zinc for several months. There may be other effects of high-dosage zinc administration for long periods of time, but these are not presently recognized.

ACKNOWLEDGMENT

I wish to thank Karen Harrington and Emiliann Quigley for their excellent help in the preparation and proofreading of this manuscript.

REFERENCES

1. **Raulin, J.,** Études cliniques sur la végétation, *Ann. Sci. Nat. Bot. Biol. Veg.,* 11, 93, 1869.
2. **Bertrand, G. and Javillier, M.,** Influence du zinc et du manganese sur la composition minerale de *L'aspergillus niger, C. R. Acad. Sci.,* 152, 1337, 1911.
3. **Sommer, A. L. and Lipman, C. B.,** Evidence on indispensable nature of zinc and boron for higher green plants, *Plant Physiol.,* 1, 231, 1926.
4. **Sommer, A. L.,** Further evidence of the essential nature of zinc for the growth of higher green plants, *Plant Physiol.,* 3, 217, 1928.
5. **Todd, W. R., Elvehjem, C. A., and Hart, E. B.,** Zinc in the nutrition of the rat, *Am. J. Physiol.,* 107, 146, 1934.
6. **Bertrand, G. and Benson, R.,** Recherches sur l'importance du zinc dans l'alimentation des animaux, *C. R., Acad. Sci.,* 175, 289, 1922.
7. **McHargue, J. S.,** Further evidence that small quantities of copper, manganese and zinc are factors in metabolism of animals, *Am. J. Physiol.,* 77, 245, 1926.
8. **Hubbell, R. B. and Mendel, L. B.,** Zinc and normal nutrition, *J. Biol. Chem.,* 75, 567, 1927.
9. **Tucker, H. F. and Salmon, W. D.,** Parakeratosis or zinc deficiency disease in pigs, *Proc. Soc. Exp. Biol. Med.,* 88, 613, 1955.
10. **O'Dell, B. L. and Savage, J. E.,** Potassium, zinc and distillers dried solubles as supplement to a purified diet, *Poult. Sci.,* 36, 459, 1957.
11. **O'Dell, B. L., Newberne, P. M., and Savage, J. E.,** Significance of dietary zinc for the growing chicken, *J. Nutr.,* 65, 503, 1958.
12. **Nishimura, H.,** Zinc deficiency in suckling mice deprived of colostrum, *J. Nutr.,* 49, 79, 1953.

13. Miller, J. K. and Miller, W. J., Development of zinc deficiency in holstein calves fed a purified diet, *J. Dairy Sci.,* 43, 1854, 1960.
14. Miller, J. K. and Miller, W. J., Experimental zinc deficiency and recovery of calves, *J. Nutr.,* 76, 467, 1962.
15. Blamberg, D. L., Blackwood, U. B., Supplee, W. C., and Combs, G. F., Effect of zinc deficiency in hens on hatchability and embryonic development, *Proc. Soc. Exp. Biol. Med.,* 104, 217, 1960.
16. Robertson, B. T. and Burns, M. J., Zinc metabolism and the zinc deficiency syndrome in the dog, *Am. J. Vet. Res.,* 24, 997, 1963.
17. Fox, M. R. S. and Harrison, B. N., Use of Japanese quail for the study of zinc deficiency, *Proc. Soc. Exp. Biol. Med.,* 116, 256, 1964.
18. Prasad, A. S., Halsted, J. A., and Nadimi, M., Syndrome of iron deficiency anemia, hepatosplenomegaly, hypogonadism, dwarfism and geophagia, *Am. J. Med.,* 31, 532, 1961.
19. Prasad, A. S., Miale, A., Jr., Farid, Z., Sandstead, H. H., and Darby, W. J., Biochemical studies on dwarfism, hypogonadism and anemia, *Arch. Intern. Med.,* 111, 407, 1963.
20. Minnich, V., Okevogla, A., Tarcon, Y., Arcasoy, A., Yorukoglu, O., Renda, F., and Demirag, B., The effect of clay on iron absorption as a possible cause for anemia of Turkish subjects with pica, *Am. J. Clin. Nutr.,* 21, 78, 1968.
21. Lemann, I. I., A study of the type of infantilism in hookworm disease, *Arch. Intern. Med.,* 6, 139, 1910.
22. Reimann, R., Wachstumsanomalien und missbildungen bei Eisenmangel-zustanden (Asiderosen), Proc. 5th Kongr. Eur. Gese. Haematol., 1955, 546.
23. Sandstead, H. H., Prasad, A. S., Schulert, A. R., Farid, Z., Miale, A., Jr., Bassilly, S., and Darby, W. J., Human zinc deficiency, endocrine manifestations and response to treatment, *Am. J. Clin. Nutr.,* 20, 422, 1967.
24. Beutler, E., Tissue effects of iron deficiency, in *Iron Metabolism,* Gross, F., Ed., Springer-Verlag, Berlin, 1964, 256.
25. Undritz, E., Oral treatment of iron deficiency, in *Iron Metabolism,* Gross, F., Ed., Springer-Verlag, Berlin, 1964, 406.
26. Darby, W. J., The oral manifestations of iron deficiency, *JAMA,* 130, 830, 1946.
27. Howell, J. T. and Monto, R. W., Syndrome of anemia, dysphagia and glossitis (Plummer-Vinson syndrome), *N. Engl. J. Med.,* 249, 1009, 1953.
28. Morrison, L. M., Swalm, W. A., and Jackson, C. L., Syndrome of hypochromic anemia, achlorhydria and atrophic gastritis, *JAMA,* 109, 108, 1937.
29. Jalili, M. A. and Al-Kassab, S., Koilonychia and cystine content of nails, *Lancet,* 2, 108, 1959.
30. Follis, R. H., Day, H. G., and McCollum, E. V., Histologic studies of the tissue of rats fed a diet extremely low in zinc, *J. Nutr.,* 22, 223, 1941.
31. Luecke, R. W., Hoefer, J. A., Brammell, W. S., and Thorp, F., Jr., Mineral interrelationships in parakeratosis in swine, *J. Anim. Sci.,* 15, 347, 1956.
32. Luecke, R. W., Hoefer, J. A., Brammell, W. S., and Schmidt, D. A., Calcium and zinc in parakeratosis in swine, *J. Anim. Sci.,* 16, 3, 1957.
33. Prasad, A. S., Sandstead, H. H., Schulert, A. R., and El Rooby, A. S., Urinary excretion of zinc in patients with the syndrome of anemia, hepatosplenomegaly, dwarfism and hypogonadism, *J. Lab. Clin. Med.,* 62, 591, 1963.
34. Vallee, B. L., Wacker, W. E. C., Bartholomay, A. F., and Hoch, F. L., Zinc metabolism in hepatic dysfunction, *N. Engl. J. Med.,* 257, 1055, 1957.
35. Prasad, A. S., Miale, A., Farid, Z., Schulert, A., and Sandstead, H. H., Zinc metabolism in patients with the syndrome of iron deficiency anemia, hypogonadism and dwarfism, *J. Lab. Clin. Med.,* 61, 531, 1963.
36. Raben, M. S., Growth hormone. I. Physiologic aspects, *N. Engl. J. Med.,* 266, 31, 1963.
37. Oberleas, D., Muhrer, M. E., and O'Dell, B. L., Some effects of phytic acid on zinc availability and physiology of swine, *J. Anim. Sci.,* 21, 57, 1962.
38. Mahdi, M. A. H. and Bassily, M., Endocrine imbalance in hepatosplenic bilharziasis, *Gaz. Kasr El Aini. Fac. Med.,* 21, 71, 1955.
39. Huang, M. H., Chang, S. C., Lu, C. W., Yu, K. J., P'an, J. P., P'an, J. S., and Kuo, P., Schistosomiasis dwarfism *China Med. J.,* 75, 449, 1957.
40. Nagaty, H. F. and Rifaat, M. A., A parasitological survey of the Kharga and of the Dakhla oases in 1955, *J. Egypt. Med. Assoc.,* 40, 444, 1957.
41. Prasad, A. S., Schulert, A. R., Miale, A., Farid, Z., and Sandstead, H. H., Zinc and iron deficiencies in male subjects with dwarfism and hypogonadism but without ancyclostomiasis, schistosomiasis or severe anemia, *Am. J. Clin. Nutr.,* 12, 437, 1963.
42. Coble, Y. D., Schulert, A. R., and Farid, Z., Growth and sexual development of male subjects in an Egyptian oasis, *Am. J. Clin. Nutr.,* 18, 421, 1966.
42a. Coble, Y. D., Van Reen, R., Schulert, A. R., Koshakji, R. P., Farid, Z., and Davis, J. T., Zinc levels and blood enzyme activities in Egyptian male subjects with retarded growth and sexual development, *Am. J. Clin. Nutr.,* 19, 415, 1966.
43. Carter, J. P., Grivetti, L. E., Davis, J. T., Nasiff, S., Mansouri, A., Mousa, W. A., Atta, A., Patwardhan, V. N., Moneim, M. A., Abdou, I. A., and Darby, W. J., Growth and sexual development of adolescent Egyptian village boys. Effects of zinc, iron, and placebo supplementation, *Am. J. Clin. Nutr.,* 22, 59, 1969.
44. Ronaghy, H., Fox, M. R. S., Garn, S. M., Israel, H., Harp, A., Moe, P. G., and Halsted, J. A., Controlled zinc supplementation for malnourished school boys: a pilot experiment, *Am. J. Clin. Nutr.,* 22, 1279, 1969.

45. **Ronaghy, H. A., Reinhold, J. G., Mahloudji, M., Ghavami, P., Fox, M. R. S., and Halsted, J. A.,** Zinc supplementation of malnourished schoolboys in Iran: increased growth and other effects, *Am. J. Clin. Nutr.,* 27, 112, 1974.
46. **Halsted, J. A., Ronaghy, H. A., Abadi, P., Haghshenass, M., Amirhakimi, G. H., Barakat, R. M., and Reinhold, J. G.,** Zinc deficiency in man: The Shiraz experiment, *Am. J. Med.,* 53, 277, 1972.
47. **Halsted, J. A., Smith, J. C., Jr., and Irwin, M. I.,** A conspectus of research on zinc requirements of man, *J. Nutr.,* 104, 345, 1974.
48. **Eminians, J., Reinhold, J. G., Kfoury, G. A., Amirhakimi, G. H., Sharif, H., and Ziai, M.,** Zinc nutrition of children in Fars Province of Iran, *Am. J. Clin. Nutr.,* 20, 734, 1967.
49. **Sandstead, H. H., Shukry, A. S., Prasad, A. S., Gabr, M. K., Hifney, A. E., Mokhtar, N., and Darby, W. J.,** Kwashiorkor in Egypt. I. Clinical and biochemical studies, with special reference to plasma zinc and serum lactic dehydrogenase, *Am. J. Clin. Nutr.,* 17, 15, 1965.
50. **Smit, Z. M. and Pretorius, P. J.,** Studies in metabolism of zinc. II. Serum zinc levels and urinary zinc excretions in South African Bantu kwashiorkor patients, *J. Trop. Pediatr.,* 9, 105, 1964.
51. **Hansen, J. D. L. and Lehmann, B. H.,** Serum zinc and copper concentrations in children with protein-calorie malnutrition, *S. Afr. Med. J.,* 43, 1248, 1969.
52. **Kumar, S. and Aro, K. S. J.,** Plasma and erythrocyte zinc levels in protein-calorie malnutrition, *Nutr. Metab.,* 15, 364, 1973.
53. **Hellwege, H. H.,** Der serumzinkspiegel und seine Veränderungen bei einigen Krankheiten im Kindesalter, *Monatsschr. Kinderheilkd.,* 119, 37, 1971.
54. **MacMahon, R. A., Parker, M. L., and McKinnon, M. C.,** Zinc treatment in malabsorption, *Med. J. Aust.,* 2, 210, 1968.
55. **Halsted, J. A. and Smith, J. C., Jr.,** Plasma-zinc in health and disease, *Lancet,* 1, 322, 1970.
56. **Solomons, N. W., Khactu, K. V., Sandstead, H. H., and Rosenberg, I. H.,** Zinc nutrition in regional enteritis (RE), *Am. J. Clin. Nutr.,* 27, 438, 1974.
57. **Moynahan, E. J. and Barnes, P. M.,** Zinc deficiency and a synthetic diet for lactose intolerance, *Lancet,* 1, 676, 1973.
58. **Green, H. L., Hambidge, K. M., and Herman, Y. F.,** Trace elements and vitamins, in *Parenteral Nutrition in Infancy and Childhood,* Winters, R., Ed., Plenum Press, New York, 1974, 131.
59. **Hambidge, K. M.,** Zinc deficiency in children, in *Proc. 2nd Int. Symp., Trace Element Metabolism in Animals,* University Park Press, New York, 1974, 322.
60. **Hambidge, K. M., Hambidge, C., Jacobs, M., and Baum, J. D.,** Low levels of zinc in hair, anorexia, poor growth, and hypogeusia in children, *Pediatr. Res.,* 6, 868, 1972.
61. **Schechter, P. J., Friedewald, W. T., Bronzert, D. A., Raff, M. S., and Henkin, R. I.,** Idiopathic hypogeusia: A description of the syndrome and a single-blind study with zinc sulfate, *Int. Rev. Neurobiol. Suppl.,* 1, 125, 1972.
62. **Sandstead, H. H.,** Zinc nutrition in the United States, *Am. J. Clin. Nutr.,* 26, 1251, 1973.
63. **Smith, J. C., Jr., McDaniel, E. G., Fan, F. F., and Halsted, J. A.,** Zinc: a trace element essential in vitamin A metabolism, *Science,* 131, 954, 1973.
64. **Henkin, R. I., Schulman, J. D., Schulman, C. B., and Bronzert, D. A.,** Changes in total, nondiffusible, and diffusible plasma zinc and copper during infancy, *J. Pediatr.,,* 82, 831, 1973.
65. **Strain, W. H., Steadman, L. T., Lankau, C. A., Jr., Berliner, W. P., and Pories, W. J.,** Analysis of zinc levels in hair for the diagnosis of zinc deficiency in man, *J. Lab. Clin. Med.,* 68, 244, 1966.
66. **Strain, W. H., Lascari, A., and Pories, W. J.,** Zinc deficiency in babies, Proc. 7th Int. Congr. Nutrition, 1966, 759.
67. **Hambidge, K. M., Walravens, P. A., Kumar, V., and Tuchinda, C.,** Chromium, zinc, manganese, copper, nickel, iron and cadmium concentrations in the hair of residents of Chandigarh, India and Bangkok, Thailand, in Trace Substances in Environmental Health, Proc. 8th Annu. Conf., Hemphill, D., Ed., University of Missouri, 1974, 39.
68. **Berfenstam, R.,** Studies on blood zinc, *Acta Paediatr.* (Stockholm), 41, 5, 1952.
69. **Bergmann, K. E. and Fomon, S. J.,** Trace minerals, in *Infant Nutrition,* Fomon, S. J., Ed., W. B. Saunders, Philadelphia, 1974, 320.
70. **Cavell, P. A. and Widdowson, E. M.,** Intakes and excretions of iron, copper, and zinc in neonatal period, *Arch. Dis. Child.,* 39, 496, 1964.
70a. *Recommended Dietary Allowances,* 8th ed., Food and Nutrition Board, National Research Council, National Academy of Sciences, Washington, D.C., 1974.
71. **Hambidge, K. M. and Silverman, A.,** Pica with rapid improvement after dietary zinc supplementation, *Arch. Dis. Child.,* 48, 567, 1973.
72. **Prasad, A. S.,** Metabolism of zinc and its deficiency in human subjects, in *Zinc Metabolism,* Prasad, A. S., Ed., Charles C Thomas, Springfield, Ill., 1966, chap. 15.
73. **Hambidge, K. M. and Walravens, P. A.,** Zinc deficiency in infants and pre-adolescent children, in *Trace Elements in Human Health and Disease,* Vol. 1, Prasad, A. S., Ed., Academic Press, New York, 1976, 21.
74. **Sandstead, H. H., Vo-Khactu, K. P., and Solomon, N.,** Conditioned zinc deficiencies, in *Trace Elements in Human Health and Disease,* Vol. 1, Prasad, A. S., Ed., Academic Press, New York, 1976, 33.

75. **Pories, W. J., Mansouri, E. G., Plecha, F. R., Flynn, A., and Strain, W. H.,** Metabolic factors affecting zinc metabolism in the surgical patient, in *Trace Elements in Human Health and Disease,* Vol. 1, Prasad, A. S., Ed., Academic Press, New York, 1976, 115.

76. **Spencer, H., Osis, D., Kramer, L., and Norris, C.,** Intake, excretion, and retention of zinc in man, in *Trace Elements in Human Health and Disease,* Vol. 1, Prasad, A. S., Ed., Academic Press, New York, 1976, 345.

77. **Reinhold, J. G., Hedayati, H., Lahimgarzadeh, A., and Nasr, K.,** Zinc, calcium, phosphorus, and nitrogen balances of Iranian villagers following a change from phytate-rich to phytate-poor diets, *Ecol. Food Nutr.,* 2, 157, 1973.

78. **Reinhold, J. G., Nasr, K., Lahimgarzadeh, A., and Hedayati, H.,** Effects of purified phytate and phytate-rich bread upon metabolism of zinc, calcium, phosphorus, and nitrogen in man, *Lancet,* 1, 283, 1973.

79. **Hsu, J. M. and Anthony, W. L.,** Zinc deficiency and urinary excretion of taurine-^{35}S and inorganic sulfate-^{35}S following cystine-^{35}S injection in rats, *J. Nutr.,* 100, 1189, 1970.

80. **Somers, M. and Underwood, E. J.,** Ribonuclease activity and nucleic acid and protein metabolism in the testes of zinc-deficient rats, *Aust. J. Biol. Sci.,* 22, 1277, 1969.

81. **Prasad, A. S., Abbasi, A., and Ortega, J.,** Zinc deficiency in man: studies in sickle cell disease in *Zinc Metabolism: Current Aspects in Health and Disease,* Brewer, G. J. and Prasad, A. S., Eds., A. R. Liss, New York, in press.

82. **Oberleas, D., Muhrer, M. E., and O'Dell, B. L.,** Dietary metal complexing agents and zinc availability in the rat, *J. Nutr.,* 90, 56, 1966.

83. **Reinhold, J. G., Faradji, B., Abadi, P., and Ismail-Beigi, F.,** Binding of zinc to fiber and other solids of wholemeal bread; with a preliminary examination of the effects of cellulose consumption upon the metabolism of zinc, calcium and phosphorus in man, in *Trace Elements in Human Health and Disease,* Vol. 1, Prasad, A. S., Ed., Academic Press, New York, 1976, 163.

84. **Reinhold, J. G.,** High phytate content of rural Iranian bread: A possible cause of human zinc deficiency, *Am. J. Clin. Nutr.,* 24, 1204, 1971.

85. **Stand, F., Rosoff, B., Williams, G. L., and Spencer, H.,** Tissue distribution studies of ionic and chelated 65-zinc in mice, *J. Pharmacol. Exp. Ther.,* 138, 399, 1962.

86. **Vohra, P. and Kratzer, F. H.,** Influence of various chelating agents on the availability of zinc, *J. Nutr.,* 82, 249, 1964.

87. **Helwig, H. L., Hoffer, E. M., Thielen, W. C., Alcocer, A. E., Hotelling, D. R., Rogers, W. H., and Lench, J.,** Urinary and serum zinc levels in chronic alcoholism, *Am. J. Clin. Pathol.,* 45, 156, 1966.

88. **Sullivan, J. F.,** Effect of alcohol on urinary zinc excretion, *Q. J. Stud. Alcohol,* 23, 216, 1962.

89. **Sullivan, J. F.,** The relation of zincuria to water and electrolyte excretion in patients with hepatic cirrhosis, *Gastroenterology,* 42, 439, 1962.

90. **Gudbjarnason, S. and Prasad, A. S.,** Cardiac metabolism in experimental alcoholism, in *Biochemical and Clinical Aspects of Alcohol Metabolism,* Sardesai, V. M., Ed., Charles C Thomas, Springfield, Ill., 1969, 266.

91. **Vallee, B. L., Wacker, W. E. C., Bartholomay, A. F., and Robin, E. D.,** Zinc metabolism in hepatic dysfunction. I. Serum zinc concentrations in Laënnec's cirrhosis and their validation by sequential analysis, *N. Engl. J. Med.,* 255, 403, 1956.

92. **Van Peenen, H. J. and Lucas, F. V.,** Zinc in liver disease, *Arch. Pathol.,* 72, 700, 1961.

93. **Van Peenen, H. J. and Patel, A.,** Tissue zinc and calcium in chronic disease, *Arch. Pathol.,* 77, 53, 1964.

94. **Sullivan, J. F. and Lankford, H. G.,** Urinary excretion of zinc in alcoholism and post alcoholic cirrhosis, *Am. J. Clin. Nutr.,* 10, 153, 1962.

95. **Sullivan, J. F. and Lankford, H. G.,** Zinc metabolism and chronic alcoholism, *Am. J. Clin. Nutr.,* 17, 57, 1965.

96. **Sullivan, J. F., Parker, M. M., and Boyett, J. F.,** Incidence of low serum zinc in noncirrhotic patients, *Proc. Soc. Exp. Biol. Med.,* 130, 591, 1969.

97. **Sullivan, J. F. and Heaney, R. P.,** Zinc metabolism in alcoholic liver disease, *Am. J. Clin. Nutr.,* 23, 170, 1970.

98. **Kahn, A. M., Helwig, H. L., Redecker, A. G., and Reynolds, T. B.,** Urine and serum zinc abnormalities in disease of the liver, *Am. J. Clin. Pathol.,* 44, 426, 1965.

99. **Prasad, A. S., Oberleas, D., and Halsted, J. A.,** Determination of zinc in biological fluids by atomic absorption spectrophotometry in normal and cirrhotic subjects, *J. Lab. Clin. Med.,* 66, 508, 1965.

100. **Halsted, J. A., Hackley, B., Rudzki, C., and Smith, J. C., Jr.,** Plasma zinc concentrations in liver diseases, *Gastroenterology,* 54, 1098, 1968.

101. **Sinha, S. N. and Gabrielli, E. R.,** Serum copper and zinc levels in various pathological conditions, *Am. J. Clin. Pathol.,* 54, 570, 1970.

102. **Boyett, J. D. and Sullivan, J. F.,** Zinc and collagen content of cirrhotic liver, *Am. J. Dig. Dis.,* 15, 797, 1970.

103. **Henkin, R. I. and Smith, F. R.,** Zinc and copper metabolism in acute viral hepatitis, *Am. J. Med. Sci.,* 264, 401, 1972.

104. **Beisel, W. R. and Pekarek, R. S.,** Acute stress and trace element metabolism, in *Neurobiology of the Trace Metals Zinc and Copper,* Pfeiffer, C. C., Ed., Academic Press, New York, 1972, 53.

105. **Patek, A. J. and Haig, C.,** The occurrence of abnormal dark adaption and its relation to vitamin A metabolism in patients with cirrhosis of the liver, *J. Clin. Invest.,* 18, 609, 1939.

106. **Kahn, A. M. and Ozeran, R.,** Liver and serum zinc abnormalities in rats and cirrhosis, *Gastroenterology,* 53, 193, 1967.

107. **Kahn, A. M., Rizer, J. G., Ponchita, T. B., and Gordon, E. H.,** Metabolism of zinc-65 in cirrhosis, *Surgery*, 63, 678, 1968.

108. **Voigt, G. E. and Saldeen, T.,** Uber den schutzeffekt des zinks gegenüber mangansulfat- oder kohlenstofftetrachlor-idinduzierten leberschäden, *Frankf. Z. Pathol.*, 74, 572, 1965.

109. **Saldeen, T. and Brunk, U.,** Enzyme histochemical investigations on the inhibitory effect of zinc on the injurious action of carbon tetrachloride on the liver, *Frankf. Z. Pathol.*, 76, 419, 1967.

110. **Saldeen, T.,** On the protective action of zinc against experimental liver damage due to choline free diet or carbon tetrachloride, *Z. Gesamte Exp. Med.*, 150, 251, 1969.

111. **Srinivasan, S. and Balwani, J. H.,** Effect of zinc sulfate on carbon tetrachloride hepatoxicity, *Acta Pharmacol. Toxicol.*, 27, 424, 1969.

112. **Chvapil, M., Ryan, J. N., and Zukowski, C. F.,** The effect of zinc on lipid peroxidation in liver microsomes and mitochondria, *Proc. Soc. Exp. Biol. Med.*, 141, 150, 1972.

113. **Davidson, C. S. and Gabuzda, G. J.,** Hepatic coma, in *Diseases of the Liver,* Schiff, L., Ed., J. B. Lippincott, Philadelphia, 1969, 378.

114. **Anthony, W. L., Woosley, R. L., and Hsu, J. M.,** Urinary excretion of radiosulfur following taurine-^{35}S injection in zinc-deficient rats, *Proc. Soc. Exp. Biol. Med.*, 138, 989, 1971.

115. **Hsu, J. M. and Woosley, R. L.,** Metabolism of L-methionine-^{35}S in zinc-deficient rats, *J. Nutr.*, 102, 1181, 1972.

116. **Zieve, L., Doizaki, W. D., and Zieve, F. J.,** Synergism between mercaptans and ammonia or fatty acids in the production of coma: A possible role of mercaptans in the pathogenesis of hepatic coma, *J. Lab. Clin. Med.*, 83, 16, 1974.

117. **Evans, G. W.,** Absorption and transport of zinc, in *Trace Elements in Human Health and Disease,* Vol. 1, Prasad, A. S., Ed., Academic Press, New York, 1976, 181.

118. **Kowerski, S., Blair-Stanek, C. S., and Schachter, D.,** Active transport of zinc and identification of zinc-binding protein in rat jejunal mucosa, *Am. J. Physiol.*, 226, 401, 1974.

119. **Vallee, B. L., Wacker, W. E. C., Bartholomay, A. F., and Hoch, F. L.,** Zinc metabolism in hepatic dysfunction, *Ann. Intern. Med.*, 50, 1077, 1959.

120. **Prasad, A. S., Schoomaker, E. B., Ortega, J., Brewer, G. J., Oberleas, D., and Oelshlegel, F. J.,** Zinc deficiency in sickle cell disease, *Clin. Chem.* (Winston-Salem), 21, 582, 1975.

121. **Szadkowski, D., Weimershaus, E., Linder, K., Schaller, K. H., and Lehnert, G.,** Einfluss von aldosteron auf den interflux und efflux arbeitsmedizinisch relevanter mineralien und spurenelemente des menschlichen organismus, *Int. Z. Angew. Physiol Einschl. Arbeitsphysiol.*, 27, 99, 1969.

122. **Schlage, C. and Worberg, B.,** Zinc in the diet of healthy preschool and school children, *Acta Paediatr.* (Stockholm), 61, 421, 1972.

123. **Osis, D., Kramer, L., Wiatrowski, E., and Spencer, H.,** Dietary zinc intake of man, *Am. J. Clin. Nutr.*, 25, 582, 1972.

124. **Underwood, E. J.,** *Trace Elements in Human and Animal Nutrition,* 3rd ed., Academic Press, New York, 1971, 208.

125. **Rose, G. A. and Willden, E. G.,** Whole blood, red cell, and plasma total and ultrafilterable zinc levels in normal subjects and in patients with chronic renal failure with and without hemodialysis, *Br. J. Urol.*, 44, 281, 1972.

126. **Mansouri, K., Halsted, J., and Gombos, E. A.,** Zinc, copper, magnesium, and calcium in dialysed and non-dialysed uremic patients, *Arch. Intern. Med.*, 125, 88, 1970.

127. **Hurley, L. S. and Tao, S.,** Alleviation of teratogenic effects of zinc deficiency by simultaneous lack of calcium, *Am. J. Physiol.*, 222, 322, 1972.

128. **Heth, D. A., Becker, W. M., and Hoekstra, W. G.,** Effect of calcium, phosphorus, and zinc on zinc-65 absorption and turnover in rats fed semi-purified diets, *J. Nutr.*, 88, 331, 1966.

129. **Haumont, S.,** Distribution of zinc in bone tissue, *J. Histochem. Cytochem.*, 9, 141, 1961.

130. **Schroeder, H. A.,** Losses of vitamins and trace minerals resulting from processing and preservation of foods, *Am. J. Clin. Nutr.*, 24, 562, 1971.

131. **Davies, I. J. T.,** Plasma-zinc concentration in patients with bronchogenic carcinoma, *Lancet*, 1, 149, 1972.

132. **Gul'ko, I, S.,** The content of zinc, copper, manganese, cadmium, cobalt, and nickel in the blood, organs, and tumors of cancer patients, *Vopr. Onkol.*, 7(9), 46, 1961.

133. **Henzel, J. H., DeWeese, M. S., and Lichti, E. L.,** Zinc concentrations within healing wounds, *Arch. Surg.* (Chicago), 100, 349, 1970.

134. **Pekarek, R. S., Wannemacher, R. W., and Beisel, W. R.,** The effect of leukocyte endogenous mediator (LEM) on the tissue distribution of zinc and iron, *Proc. Soc. Exp. Biol. Med.*, 140, 685, 1972.

135. **Lindeman, R. D., Bottomley, R. G., Cornelison, R. L., and Jacobs, L. A.,** Influence of acute tissue injury on zinc metabolism in man, *J. Lab. Clin. Med.*, 79, 452, 1972.

136. **Cohen, I. K., Schechter, P. J., and Henkin, R. I.,** Hypogeusia, anorexia, and altered zinc metabolism following thermal burn, *J.A.M.A.*, 223, 914, 1973.

137. **Nielsen, J. P. and Jemec, B.,** Zinc metabolism in patients with severe burns, *Scand. J. Plast. Reconstr. Surg.*, 2, 47, 1968.

138. **Cuthbertson, D. P., Fell, G. S., Smith, C. M., and Tolstone, W. J.,** Metabolism after injury. I. Effects of severity, nutrition, and environmental temperature on protein, potassium, zinc, and creatine, *Br. J. Surg.*, 59, 925, 1972.

139. **Sandstead, H. H., Glasser, S. R., and Gillespie, D. D.,** Zinc deficiency: Effect on fetal growth, zinc concentration, and 65-zinc uptake, *Fed. Proc. Fed. Am. Soc. Exp. Biol.,* 29, 297, 1970.
140. **Husain, S. L.,** Oral zinc sulfate in leg ulcers, *Lancet,* 1, 1069, 1969.
141. **Serjeant, G. R., Galloway, R. E., and Gueri, M. C.,** Oral zinc and sulphate in sickle-cell ulcers, *Lancet,* 2, 891, 1970.
142. **Greaves, M. W.,** Zinc in cutaneous ulceration due to vascular insufficiency, *Am. Heart J.,* 83, 716, 1972.
143. **Haeger, K., Lanner, E., and Magnusson, P. O.,** Oral zinc sulfate in the treatment of venous leg ulcers, *Vasa,* 1, 62, 1972.
144. **Oberleas, D., Seymour, J. K., Lenaghan, R., Hovanesian, J., Wilson, R. F., and Prasad, A. S.,** Effect of zinc deficiency on wound healing in rats, *Am. J. Surg.,* 121, 566, 1971.
145. **Fernandez-Madrid, F., Prasad, A. S., and Oberleas, D.,** Effect of zinc deficiency on nucleic acids, collagen, and noncollagenous protein of the connective tissue, *J. Lab. Clin. Med.,* 82, 951, 1973.
146. **Greaves, M. W. and Boyde, T. R. C.,** Plasma zinc concentrations in patients with psoriasis, other dermatosis, and venous ulcerations, *Lancet,* 2, 1019, 1967.
147. **Greaves, M. W.,** Zinc and copper in psoriasis, *Br. J. Dermatol.,* 86, 439, 1972.
148. **Portnoy, B. and Molokhia, M. M.,** Zinc and copper in psoriasis, *Br. J. Dermatol.,* 85, 597, 1971.
149. **Portnoy, B. and Molokhia, M. M.,** Zinc and copper in psoriasis, *Br. J. Dermatol.,* 86, 205, 1972.
150. **Pories, W. J., Henzel, J. H., Rob, C. G., and Strain, W. H.,** Acceleration of healing with zinc sulfate, *Ann. Surg.,* 165, 432, 1967.
151. **Pories, W. J., Henzel, J. H., Rob, C. G., and Strain, W. H.,** Acceleration of wound healing in man with zinc sulfate given by mouth, *Lancet,* 1, 121, 1967.
152. **Pories, W. J. and Strain, W. H.,** Zinc and wound healing, in *Zinc Metabolism,* Prasad, A. S., Ed., Charles C Thomas, Springfield, Ill., 1966, 378.
152a. **Sunderman, F. W., Jr.,** Current status of zinc deficiency in the pathogenesis of neurological, dermatological, and musculosketal disorders. *Ann. Clin. Lab. Sci.,* 5, 132, 1975.
153. **Cohen, C.,** Zinc sulfate and bedsores, *Br. Med. J.,* 2, 561, 1968.
154. **Greaves, M. W. and Skillen, A. W.,** Effects of long-continued ingestion of zinc sulphate in patients with venous leg ulceration, *Lancet,* 2, 889, 1970.
155. **Barcia, P. J.,** Lack of acceleration of healing with zinc sulfate, *Ann. Surg.,* 172, 1048, 1970.
156. **Myers, M. B. and Cherry, G.,** Zinc and the healing of chronic leg ulcers, *Am. J. Surg.,* 120, 77, 1970.
157. **Clayton, R. J.,** Double-blind trial of oral zinc sulphate in patients with leg ulcers, *Br. J. Clin. Pract.,* 26, 368, 1972.
158. **Greaves, M. W. and Ive, F. A.,** Double-blind trial of zinc sulphate in the treatment of chronic venous leg ulceration, *Br. J. Dermatol.,* 87, 632, 1972.
159. **Hallböök, T. and Lanner, E.,** Serum zinc and healing of venous leg ulcers, *Lancet,* 2, 780, 1972.
160. **Husain, S. L., Fell, G. S., and Scott, R.,** Zinc and healing, *Lancet,* 2, 1361, 1970.
161. **Rahmat, A., Norman, J. N., and Smith, G.,** The effect of zinc deficiency on wound healing, *Br. J. Surg.,* 61, 271, 1974.
162. **Sandstead, H. H., Burk, R. F., Booth, G. H., and Darby, W. J.,** Current concepts on trace minerals: clinical considerations, *Med. Clin. North Am.,* 54, 1509, 1970.
163. **Sandstead, H. H. and Shepard, G. H.,** The effect of zinc deficiency on the tensile strength of healing surgical incisions in the integument of the rat, *Proc. Soc. Exp. Biol. Med.,* 128, 687, 1968.
164. **McClain, P. E., Wiley, E. R., Beecher, G. R., Anthony, W. L., and Hsu, J. M.,** Influence of zinc deficiency on synthesis and cross-linking of rat-skin collagen, *Biochim. Biophys. Acta,* 304, 457, 1973.
165. **Stephen, J. K. and Hsu, J. M.,** Effect of zinc deficiency and wounding on DNA synthesis in the rat skin, *J. Nutr.,* 103, 548, 1973.
166. **Groundwater, W. and MacLeod, I. B.,** The effects of systemic zinc supplements on the strength of healing incised wounds in normal rats, *Br. J. Surg.,* 57, 222, 1970.
167. **Lee, P. W. R. and Green, M. A.,** Zinc and wound healing, *Lancet,* 2, 1089, 1972.
168. **Elias, S. and Chvapil, M.,** Zinc and wound healing in normal and chronically ill rats, *J. Surg. Res.,* 15, 59, 1973.
169. **Flynn, A., Pories, W. J., Strain, W. H., Hill, O. A., and Fratianne, R. B.,** Rapid serum-zinc depletion associated with corticosteroid therapy, *Lancet,* 2, 1169, 1971.
170. **Klingberg, W. G., Prasad, A. S., and Oberleas, D.,** Zinc deficiency following Penicillamine therapy, in *Trace Elements in Human Health and Disease,* Vol. 1, Prasad, A. S., Ed., Academic Press, New York, 1976, 51.
171. **Tarui, S.,** Studies on zinc metabolism. III. Effects of the diabetic state on zinc metabolism. A clinical aspect, *Endocrinol. Jpn.,* 10, 9, 1963.
172. **Pidduck, H. G., Wren, P. J. J., and Price Evans, D. A.,** Plasma zinc and copper in diabetes mellitus, *Diabetes,* 19, 234, 1970.
173. **McCall, J. T., Goldstein, N. P., and Smith, L. H.,** Implications of trace metals in human disease, *Fed. Proc. Fed. Am. Soc. Exp. Biol.,* 30, 1011, 1971.
174. **Henkin, R. I., Marshall, J. R., and Meret, S.,** Maternal-fetal metabolism of copper and zinc at term, *Am. J. Obstet. Gynecol.,* 110, 131, 1971.
175. **O'Dell, B. L., Burpo, C. E., and Savage, J. E.,** Evaluation of zinc availability in foodstuffs of plant and animal origin, *J. Nutr.,* 102, 653, 1972.

176. **Hurley, L. S.,** Perinatal effects of trace element deficiencies, in *Trace Elements in Human Health and Disease,* Vol. 2, Prasad, A. S., Ed., Academic Press, New York, 1976, 301.

177. **Caldwell, D. F., Oberleas, D., Clancy, J. J., and Prasad, A. S.,** Behavioral impairment in adult rats following acute zinc deficiency, *Proc. Soc. Exp. Biol. Med.,* 133, 1417, 1970.

178. **Halas, E. S., Rowe, M. C., Orris, R. J., McKenzie, J. M., and Sandstead, H. H.,** Effects of intra-uterine zinc deficiency on subsequent behavior, in *Trace Elements in Human Health and Disease,* Prasad, A. S., Ed., Academic Press, New York, 1976, 327.

179. **Lokken, P. M., Halas, E. S., and Sandstead, H. H.,** Influence of zinc deficiency on behavior, *Proc. Soc. Exp. Biol. Med.,* 144, 680, 1973.

180. **Hurley, L. S. and Shrader, R. E.,** Congenital malformations of the nervous system of zinc deficient rats, *Int. Rev. Neurobiol.,* Suppl. 1, 7, 1972.

181. **Damyanov, I. and Dutz, W.,** Anencephaly in Shiraz, Iran, *Lancet,* 1, 82, 1971.

182. **Schlievert, P., Johnson, W., and Galask, R. P.,** Bacterial growth inhibition by amniotic fluid. V. Phosphate to zinc ratio as a predictor of bacterial inhibitory activity, *Am. J. Obstet. Gynecol.,* in press.

183. **Schlievert, P., Johnson, W., and Galask, R. P.,** Bacterial growth inhibition by amniotic fluid. VI. Evidence for a zinc peptide. Antibacterial system, *Am. J. Obstet. Gynecol.,* in press.

184. **Prasad, A. S., Oberleas, D., Lei, K. Y., Moghissi, K. S., and Stryker, J. C.,** Effect of oral contraceptive agents on nutrients. I. Minerals, *Am. J. Clin. Nutr.,* 28, 377, 1975.

185. **Halsted, J. A., Hackley, B. M., and Smith, J. C., Jr.,** Plasma-zinc and copper in pregnancy and after oral contraceptives, *Lancet,* 2, 278, 1968.

186. **O'Leary, J. A. and Spellacy, W. N.,** Zinc and copper levels in pregnant women and those taking oral contraceptives, *Am. J. Obstet. Gynecol.,* 102, 131, 1969.

186a. **Abbasi, A. A., Prasad, A. S., Ortega, J., Congco, E., and Oberleas, D.,** Gonadal function abnormalities in sickle cell anemia: studies in adult male patients, *Ann. Int. Med.,* 85, 601, 1976.

187. **Oelshlegel, F. J., Brewer, G. J., Knutsen, C., Prasad, A. S., and Schoomaker, E. B.,** Studies on the interaction of zinc with human hemoglobin, *Arch. Biochem. Biophys.,* 163, 742, 1974.

188. **Oelshlegel, F. J., Brewer, G. J., Prasad, A. S., Knutsen, C., and Schoomaker, E. B.,** Effect of zinc on increasing oxygen affinity of sickle and normal red blood cells, *Biochem. Biophys. Res. Commun.,* 53, 560, 1973.

189. **Brewer, G. J., Oelshlegel, F. J., Jr., and Prasad, A. S.,** Zinc in sickle cell anemia, in *Erythrocyte Structure and Function,* Vol. 1, Brewer, G. J., Ed., Alan R. Liss, New York, 1975, 417.

189a. **Brewer, G. J., Brewer, L. F., and Prasad, A. S.,** Suppression of irreversibly sickled erythrocytes by zinc therapy in sickle cell anemia, *J. Lab. Clin. Med.,* in press.

190. **Danbolt, N. and Closs, K.,** Akrodermatitis enteropathica, *Acta Derm. Venereol,* 23, 127, 1942.

191. **Aguilera-Diaz, L. M.,** Un nouveau symptome dans l'acrodermatite ẽntéropathique: la démarche ataxique, *Bull. Soc. Fr. Dermatol. Syphiligr,* 78, 259, 1971.

192. **Entwisle, B. R.,** Acrodermatitis enteropathica: Report of a case in a twin with dramatic response to expressed human milk, *Aust. J. Dermatol.,* 8, 13, 1965.

193. **Frier, S., Faber, J., Goldstein, R., and Mayer, M.,** Treatment of acrodermatitis enteropathica by intravenous amino acid hydrolysate, *J. Pediatr.,* 82, 109, 1973.

194. **Juljulian, H. H. and Kurban, A. K.,** Acantholysis: a feature of acrodermatitis enteropathica, *Arch. Dermatol.,* 103, 105, 1971.

195. **Perry, H. O.,** Acrodermatitis enteropathica, in *Clinical Dermatology,* Vol. 1, Demis, D. J., Crounse, R. G., Dobson, R. L., and McGuire, J., Eds., Harper and Row, New York, 1974, 1.

196. **Rodin, A. E. and Goldman, A. S.,** Autopsy findings in acrodermatitis enteropathica, *Am. J. Clin. Pathol.,* 51, 315, 1969.

197. **Wells, B. T. and Winkelmann, R. K.,** Acrodermatitis enteropathica: Report of 6 cases, *Arch. Dermatol.,* 84, 40, 1961.

198. **Dillaha, C. J., Lorincz, A. L., and Aavick, O. R.,** Acrodermatitis enteropathica, *JAMA,* 152, 509, 1953.

199. **Moynahan, E. J.,** Acrodermatitis enteropathica with secondary lactose intolerance and tertiary deficiency state, probably due to chelation essential nutrients by diiodohydroxyquinolone, *Proc. R. Soc. Med.,* 59, 7, 1966.

200. **Barnes, P. M. and Moynahan, E. J.,** Zinc deficiency in acrodermatitis enteropathica: multiple dietary intolerance treated with synthetic diet, *Proc. R. Soc. Med.,* 66, 327, 1973.

201. **Brummerstedt, E., Flagstad, T., and Andresen, E.,** The effect of zinc on calves with hereditary thymus hypoplasia (lethal tract A 46), *Acta Pathol., Microbiol. Scand. Sect. A,* 79, 686, 1971.

201a. **Smith, J. C., Jr., Zeller, J. A., Brown, E. D., and Ong, S. C.,** Elevated plasma zinc: a heritable anomaly, *Science,* 193, 496, 1976.

201b. **Simkin, P. A.,** Oral zinc sulphate in rheumatoid arthritis, *Lancet,* 2, 539, 1976.

202. **Tipton, I. H. and Cook, M. J.,** Trace elements in human tissue. Part II. Adult subjects from the United States, *Health Phys.,* 9, 103, 1963.

202a. **Tipton, I. H., Schroeder, H. A., Perry, H. M., Jr., and Cook, M. J.,** Trace elements in human tissue. Part III. Subjects from Africa, the Near and Far East and Europe, *Health Phys.,* 11, 403, 1965.

202b. **McBean, L. D., Dove, J. T., Halsted, J. A., and Smith, J. C., Jr.,** Zinc concentration in human tissues, *Am. J. Clin. Nutr.,* 25, 672, 1972.

203. **Prasad, A. S.,** Deficiency of zinc in man and its toxicity, in *Trace Elements in Human Health and Disease,* Vol. 1, Prasad, A. S., Eds., Academic Press, New York, 1976, 1.

204. **Foley, B., Johnson, S. A., Hackley, B., Smith, J. C., Jr., and Halsted, J. A.,** Zinc content of human platelets, *Proc. Soc. Exp. Biol. Med.,* 128, 265, 1968.

205. **Prasad, A. S., Schoomaker, E. B., Ortega, J., Brewer, G. J., Oberleas, D., and Oelshlegel, F.,** Role of zinc in man and its deficiency in sickle cell disease, in *Erythrocyte Structure and Function,* Vol. 1, Brewer, G. J., Ed., Alan R. Liss, New York, 1975, 603.

206. **Prasad, A. S. and Oberleas, D.,** Binding of zinc to amino acids and serum proteins in vitro, *J. Lab. Clin. Med.,* 76, 416, 1970.

207. **Ross, J. F., Ebaugh, F. G., Jr., and Talbot, T. R., Jr.,** Radioisotopic studies of zinc metabolism in human subjects, *Trans. Assoc. Am. Physicians,* 71, 322, 1958.

208. **McBean, L. D. and Halsted, J. A.,** Fasting versus postprandial plasma zinc levels, *J. Clin. Pathol.,* 22, 623, 1969.

209. **Pearson, W. N. and Reich, M.,** *In vitro* studies of Fe^{59} absorption by everted intestinal sacs of the rat, *J. Nutr.,* 123, 117, 1965.

210. **Methfessel, A. H. and Spencer, H.,** Zinc metabolism in the rat. II. Secretion of zinc into intestines, *J. Appl. Physiol.,* 34, 63, 1973.

211. **Becker, W. M. and Hoekstra, W. G.,** The intestinal absorption of zinc, in *Intestinal Absorption of Metal Ions, Trace Elements and Radionuclides,* Skoryna, S. C. and Waldron-Edward, D., Eds., Pergamon Press, New York, 1971, 229.

212. **O'Dell, B. L. and Savage, J. E.,** Effect of phytic acid on zinc availability, *Proc. Soc. Exp. Biol. Med.,* 103, 304, 1960.

213. **Oberleas, D. and Prasad, A. S.,** Growth as affected by zinc and protein nutrition, symposium on zinc metabolism, *Am. J. Clin. Nutr.,* 22, 1304, 1969.

214. **Likuski, H. J. A. and Forbes, R. M.,** Effect of phytic acid on the availability of zinc in amino acid and casein diets fed to chicks, *J. Nutr.,* 84, 145, 1964.

215. **Scoular, F. L.,** A quantitative study, by means of spectrographic analysis, of zinc in nutrition, *J. Nutr.,* 17, 103, 1939.

216. **Macy, I. G.,** Nutrition and chemical growth in childhood, in *Evaluation,* Vol. I, Charles C Thomas, Springfield, Ill., 1942, 198.

217. **Tribble, H. M. and Scoular, F. I.,** Zinc metabolism of young college women on self-selected diets, *J. Nutr.,* 52, 209, 1954.

218. **Engel, R. W., Miller, R. F., and Price, N. O.,** Metabolic patterns in preadolescent children. XIII. Zinc balance, in *Zinc Metabolism,* Prasad, A. S., Ed., Charles C Thomas, Springfield, Ill., 1966, 326.

219. **McCance, R. A. and Widdowson, E. M.,** The absorption and excretion of zinc, *Biochem. J.,* 36, 692, 1942.

220. **Tipton, I. H., Stewart, P. L., and Dickson, J.,** Patterns of elemental excretion in long term balance studies, *Health Phys.,* 16, 455, 1969.

221. **White, H. S. and Gynee, T. N.,** Utilization of inorganic elements by young women eating iron-fortified foods, *J. Am. Diet. Assoc.,* 59, 27, 1971.

222. **Gormican, A. and Catli, E.,** Mineral balance in young men fed a fortified milk-base formula, *Nutr. Metab.,* 13, 364, 1971.

223. **Price, N. O. Bunce, G. E., and Engel, R. W.,** Copper, manganese, and zinc balance in preadolescent girls, *Am. J. Clin. Nutr.,* 23, 258, 1970.

224. **Murphy, E. W. Page, L., and Watt, B. K.,** Trace minerals in type A school lunches, *J. Am. Diet. Assoc.,* 58, 115, 1971.

225. **Spencer, H., Rosoff, B., Feldstein, A., Cohn, S. H., and Gusmano, E.,** Metabolism of zinc-65 in man, *Radiat. Res.,* 24, 432, 1965.

226. **Spencer, H., Vankinscott, V., Lewin, I., and Samachson, J.,** Zinc-65 metabolism during low and high calcium intake in man, *J. Nutr.,* 86, 169, 1965.

227. **Prasad, A. S., Schulert, A. R., Sandstead, H. H., Miale, A., and Farid, Z.,** Zinc, iron, and nitrogen content of sweat in normal and deficient subjects, *J. Lab. Clin. Med.,* 62, 84, 1963.

228. **Schroeder, H. A., Nason, A. P., Tipton, I. H., and Balassa, J. J.,** Essential trace metals in man: zinc: relation to environmental cadmium, *J. Chronic Dis.,* 20, 179, 1967.

229. **Fell, G. S., Fleck, A., Cuthbertson, D. P., Queen, K., Morrison, C., Bessent, R. G., and Husain, S. L.,** Urinary zinc levels as an indication of muscle catabolism, *Lancet,* 1, 280, 1973.

230. **Davies, J. W. L. and Fell, G. S.,** Tissue catabolism in patients with burns, *Clin. Chim. Acta,* 51, 83, 1974.

231. **Vallee, B. L.,** Biochemistry, physiology and pathology of zinc, *Physiol. Rev.,* 39, 443, 1959.

232. **Keilin, D. and Mann, J.,** Carbonic anhydrase. Purification and nature of the enzyme, *Biochem. J.,* 34, 1163, 1940.

233. **Riordan, J. F. and Vallee, B. L.,** Structure and function of zinc metalloenzymes, in *Trace Elements in Human Health and Disease,* Vol. 1, Prasad, A. S., Ed., Academic Press, New York, 1976, 227.

234. **Vallee, B. L. and Wacker, W. E. C.,** Metalloproteins, in *The Proteins: Composition, Structure and Function,* 2nd ed., Neurath, H., Ed., Academic Press, New York, 1970, 199.

235. **Vallee, B. L.,** Zinc and metalloenzymes, *Adv. Protein Chem.,* 10, 317, 1955.

236. **Rosenbusch, J. P. and Weber, K.,** Localization of the zinc binding site of aspartate transcarbamylase in the regulatory subunit, *Proc. Natl. Acad. Sci. U.S.A.,* 68, 1019, 1971.

237. **Drum, D. E., Harrison, J. H., Li, T. -K., Bethune, J. L., and Vallee, B. L.,** Structural and functional zinc in horse liver alcohol dehydrogenase, *Proc. Natl. Acad. Sci. U.S.A.,* 57, 1434, 1967.

238. **Carpenter, F. H. and Vah., J. M,.** Leucine aminopeptidase (bovine lens) mechanism of activation by Mg^{2+} and Mn^{2+} of the zinc metalloenzyme amino acid composition, and sulfhydryl content, *J. Biol. Chem.,* 248, 294, 1973.

239. **Prasad, A. S. and Oberleas, D.,** Thymidine kinase activity and incorporation of thymidine into DNA in zinc-deficient tissue, *J. Lab. Clin. Med.,* 83, 634, 1974.

240. **Prasad, A. S. and Oberleas, D.,** Changes in activity of zinc-dependent enzymes in zinc-deficient tissues of rats, *J. Appl. Physiol.,* 31, 842, 1971.

240a. **Kirchgessner, M., Roth, H. P., and Weigand, E.,** Biochemical changes in zinc deficiency, in *Trace Elements in Human Health and Disease,* Vol. 1, Prasad, A. S., Ed., Academic Press, New York, 1976.

241. **Kfoury, G. A., Reinhold, J. G., and Simonian, S. J.,** Enzyme activities in tissues of zinc-deficient rats, *J. Nutr.,* 95, 102, 1968.

242. **Macapinlac, M. P., Pearson, W. N., and Darby, W. J.,** Some characteristics of zinc deficiency in the albino rat, in *Zinc Metabolism,* Prasad, A. S., Ed., Charles C Thomas, Springfield, Ill., 1966, 142.

243. **Luecke, R. W. and Baltzer, B. V.,** The effect of dietary intake on the activity of intestinal alkaline phosphatase in the zinc-deficient rat, *Fed. Proc., Fed. Am. Soc. Exp. Biol.,* 27, 483, 1968.

244. **Luecke, R. W., Olman, M. E., and Baltzer, B. V.,** Zinc deficiency in the rat: Effect on serum and intestinal alkaline phosphatase activities, *J. Nutr.,* 94, 344, 1968.

245. **Roth, H. P. and Kirchgessner, M.,** Aktivitätsveränderungen verschiedener Dehydrogenasen und der Alkalischen Phosphatase in Serum bei zn Depletion und Repletion, *Z. Tierphysiol. Tierenaehr. Futtermittelkd.,* 32, 289, 1974.

246. **Guttikar, M. N., Panemangalore, M., and Roa, M. N.,** Effects of protein-calorie ratio on liver enzyme concentration in young weanling rats, *Nutr. Metab.,* 12, 136, 1970.

247. **Day, H. G. and McCollum, E. V.,** Effects of acute dietary zinc deficiency in the rat, *Proc. Soc. Exp. Biol. Med.,* 45, 282, 1940.

248. **Shrader, R. E. and Hurley, L. S.,** Enzyme histochemistry of peripheral blood and bone marrow in zinc-deficient rats, *Lab. Invest.,* 26, 566, 1972.

249. **Roth, H. P. and Kirchgessner, M.,** Zum einfluβ unterschliedlicher Diätzinkgehalte auf die Aktivität der Alkalischen Phosphatase im Knochen, *Z. Tierphysiol. Tierenaehr. Futtermittelkd.,* 33, 57, 1974.

250. **Roth, H. P. and Kirchgessner, M.,** De- und Repletionsstudien an Zink im Knochen un leber wachsender Ratten, *Arch. Tierernaehr.,* 24, 283, 1974.

251. **Starcher, B. and Kratzer, F. H.,** Effect of zinc on bone alkaline phosphatase in turkey poults, *J. Nutr.,* 79, 18, 1963.

252. **Davies, M. J. and Motzok, J.,** Zinc deficiency in the chick: Effect on tissue alkaline phosphatases, *Comp. Biochem. Physiol. B,* 40, 129, 1971.

253. **Iqbal, M.,** Activity of alkaline phosphatase and carbonic anhydrase in male and female zinc-deficient rats, *Enzyme,* 12, 33, 1971.

254. **Williams, R. B.,** Intestinal alkaline phosphatase and inorganic pyrophosphatase activities in the zinc-deficient rat, *Br. J. Nutr.,* 27, 121, 1972.

255. **Roth, H. P. and Kirchgessner, M.,** Zur aktivität der pankreascarboxypeptidase A and B bei zink-depletion and repletion, *Z. Tierphysiol. Tierenaehr. Futtermittelkd.* 33, 62, 1974.

256. **Hsu, J. M., Anilane, J. K., and Scanlan, D. E.,** Pancreatic carboxypeptidase: Activities in zinc-deficient rats, *Science,* 153, 882, 1966.

257. **Roth, H. P. and Kirchgessner, M.,** Zur aktivitat der blut-carbon-anhydrase bei zn-mangel wachsender ratten, *Z. Tierphysiol. Tierenaehr. Futtermittelkd.,* 32, 286, 1974.

258. **Ott, E. A., Smith, W. H., Stob, M., Parker, H. E., Harrington, R. B., and Beeson, W. M.,** Zinc requirement of the growing lamb fed a purified diet, *J. Nutr.,* 87, 459, 1965.

259. **Prasad, A. S., Oberleas, D., Wolf, P., and Horwitz, J. P.,** Studies on zinc deficiency: Changes in trace elements and enzyme activities in tissues of zinc-deficient rats, *J. Clin. Invest.,* 46, 549, 1967.

260. **Prasad, A. S., Oberleas, D., Wolf, P., Horwitz, J. P., Miller, E. R., and Luecke, R. W.,** Changes in trace elements and enzyme activities in tissues of zinc-deficient pigs, *Am. J. Clin. Nutr.,* 22, 628, 1969.

261. **Prasad, A. S., Oberleas, D., Wolf, P., and Horwitz, J. P.,** Effect of growth hormone on non-hypophysectomized zinc-deficient rats and zinc on hypophysectomized rats, *J. Lab. Clin. Med.,* 73, 486, 1969.

262. **Prasad, A. S., Oberleas, D., Miller, E. R., and Luecke, R. W.,** Biochemical effects of zinc deficiency: Changes in activities of zinc-dependent enzymes and ribonucleic acid and deoxyribonucleic acid content of tissues, *J. Lab. Clin. Med.,* 77, 144, 1971.

263. **Roth, H. P. and Kirchgessner, M.,** Zum Aktivitätsverlauf verschiedener Dehydrogenasen in der Rattenleber bei unterschleidlicher Zinkversorgung, *Z. Tierphysiol. Tierenaehr. Futtermittelkd.,* 33, 1, 1974.

264. **Huber, A. M. and Gershoff, S. N.,** Effects of zinc deficiency on the oxidation of retinol and ethanol in rats, *J. Nutr.,* 105, 1486, 1975.

265. **Roth, H. P. and Kirchgessner, M.,** Zur Enzymaktivitat von Dehydrogenasen im Rattenmuskle bei Zinkmangel, *Z. Tierphysiol. Tierenaehr. Futtermittelkd.,* 33, 67, 1974.

266. Swenerton, H., Shrader, R., and Hurley, L. S., Lactic and malic dehydrogenases in testes of zinc-deficient rats, *Proc. Soc. Exp. Biol. Med.*, 141, 283, 1972.

267. Hurley, L. S., Zinc deficiency in the developing rats, *Am. J. Clin. Nutr.*, 22, 1332, 1969.

268. Swenerton, H. and Hurley, L. S., Severe zinc deficiency in male and female rats, *J. Nutr.*, 95, 8, 1968.

269. Reinhold, J. G. and Kfoury, G. A., Zinc-dependent enzyme in zinc-depleted rats: Intestinal alkaline phosphatase, *Am. J. Clin. Nutr.*, 22, 1250, 1969.

270. Mills, C. F., Quarterman, J., Chesters, J. K., Williams, R. B., and Dalgarno, A. C., Metabolic role of zinc, *Am. J. Clin. Nutr.*, 22, 1240, 1969.

271. Prasad, A. S. and Oberleas, D., Zinc: Human nutrition and metabolic effects, *Ann. Intern. Med.*, 73, 631, 1970.

272. Kirchgessner, M. and Roth, H. P., Beziehunger Zwischen Klinischen Mangelsymptomen und Enzymaktivitäten bei Zinkmangel, *Zentralbl. Veterinaermed. Reihe A*, 22, 14, 1975.

273. Kirchgessner, M. and Roth, H. P., Zur Bestimmug der Verfügbarkeit von Zink in stoffwechsel Sowie zur Ermittlung des zinkdebarfs Mittels Aktivitäsänderunger von Zn-metalloenzymen, *Arch. Tierernaehr.*, 25, 83, 1975.

274. Kirchgessner, M., Müller, H. L., Wiegand, E., Grassmann, E., Schwarz, F. J., Pallauf, J., and Roth, H. P., Zur Definition und Bestimmug der absorbierbarkeit, intermediären Verfügbarkeit und Gesamtverwertung von essentiellen Spurenelementen, *Z. Tierphysiol. Tierenaehr. Futtermittelkd.*, 34, 3, 1974.

275. Macapinlac, M. P., Pearson, W. N., Barney, G. H., and Darby, W. J., Protein and nucleic acid metabolism in the testes of zinc-deficient rats, *J. Nutr.*, 95, 569, 1968.

276. Sandstead, H. H., Terhune, M., Brady, R. N., Gillespie, D., and Hollaway, W. L., Zinc deficiency: Brain DNA, protein and lipids; and liver ribosomes and RNA polymerase, *Clin. Res.*, 19, 83, 1971.

277. Fox, M. R. S. and Harrison, B. N., Effects of zinc deficiency on plasma proteins of young Japanese quail, *J. Nutr.*, 86, 89, 1965.

278. Miller, E. R., Luecke, R. W., Ullrey, D. E., Baltzer, B. V., Bradley, B. L., and Hoefer, J. A., Biochemical, skeletal, and allometric changes due to zinc deficiency in the baby pig, *J. Nutr.*, 95, 278, 1968.

279. Tao, S. and Hurley, L. S., Changes in plasma proteins in zinc-deficient rats, *Proc. Soc. Exp. Biol. Med.*, 136, 165, 1971.

280. Theuer, R. C. and Hoekstra, W. G., Oxidation of ^{14}C-labeled carbohydrate, fat, and amino acid substrates by zinc-deficient rats, *J. Nutr.*, 89, 448, 1966.

281. Mills, C. F., Quarterman, J., Williams, R. B., Dalgarno, A. C., and Panic, B., The effects of zinc deficiency on pancreatic carboxypeptidase activity and protein digestion and absorption in the rat, *Biochem. J.*, 102, 712, 1967.

282. Schneider, E. and Price, C. A., Decreased ribonucleic acid levels: A possible cause of growth inhibition in zinc deficiency, *Biochim. Biophys. Acta*, 55, 406, 1962.

283. Wacker, W. E. C., Nucleic acid and metals. III. Changes in nucleic acid, protein, and metal content as a consequence of zinc deficiency in *Euglena gracilis*, *Biochemistry*, 1, 859, 1962.

284. Wegener, W. S. and Romano, A. H., Control isocitratase formation in *Rhizopus nigricans*, *J. Bacteriol.*, 87, 156, 1964.

285. Kessler, B. and Monselise, S. P., Studies on ribonuclease, ribonucleic acid, and protein synthesis in healthy and zinc-deficient citrus leaves, *Physiol. Plant.*, 12, 1, 1959.

286. Ku, P. K., Ullery, D. E., and Miller, E. R., Zinc deficiency and tissue nucleic acid and protein concentration, in *Trace Element Metabolism in Animals*, Mills, C. F., Ed., E. and S. Livingstone, Edinburgh, 1970, 158.

287. Prasad, A. S. and Oberleas, D., Ribonuclease and deoxyribonuclease activities in zinc-deficient tissues, *J. Lab. Clin. Med.*, 82, 461, 1973.

288. Ohtake, Y., Uchida, K., and Sakai, T., Purification and properties of ribonuclease from yeast, *J. Biochem.* (Tokyo), 54, 322, 1963.

289. O'Neal, R. M., Pla, G. W., Fox, M. R. S., Gibson, F. S., and Fry, B. E., Effect of zinc deficiency and restricted feeding on protein and ribonucleic acid metabolism of rat brain, *J. Nutr.*, 100, 491, 1970.

290. Williams, R. B., Mills, C. F., Quarterman, J., and Dalgarno, A. C., The effect of zinc deficiency on the in vivo incorporation of ^{32}P into rat-liver nucleotides, *Biochem. J.*, 95, 29, 1965.

291. Weser, U., Seeber, S., and Warnecke, P., Reactivity of Zn^{2+} on nuclear DNA and RNA biosynthesis of regeneration in rat liver, *Biochim. Biophys. Acta*, 179, 422, 1969.

292. Weser, U., Seeber, S., and Warnecke, P., Effect of Zn^{2+} on nuclear RNA and protein-biosynthesis in rat liver, *Z. Naturforsch. Teil B*, 24, 866, 1969.

293. Rubin, H. and Koide, T., Inhibition of DNA synthesis in chick embryo cultures by deprivation of either serum or zinc, *J. Cell Biol.*, 56, 777, 1973.

294. Rubin, H., Inhibition of DNA synthesis in animal cells by ethylene diamine tetraacetate and its reversal by zinc, *Proc. Natl. Acad. Sci. U.S.A.*, 69, 712, 1972.

295. Fujioka, M. and Lieberman, I., A Zn^{++} requirement for synthesis of deoxyribonucleic acid by rat liver, *J. Biol. Chem.*, 239, 1164, 1964.

296. Sandstead, H. H. and Rinaldi, R. A., Impairment of deoxyribonucleic acid synthesis by dietary zinc deficiency in the rat, *J. Cell. Physiol.*, 73, 81, 1969.

297. Swenerton, H., Shrader, R., and Hurley, L. S., Zinc-deficient embryos: Reduces thymidine incorporation, *Science*, 166, 1014, 1969.

298. **Williams, R. B. and Chesters, J. K.,** Effects of zinc deficiency on nucleic acid synthesis in the rat, in *Trace Element Metabolism in Animals,* Mills, C. F., Ed., E. and S. Livingston, Edinburgh, 1970, 164.

299. **Williams, R. B. and Chesters, J. K.,** The effects of early Zn deficiency on DNA and protein synthesis in the rat, *Br. J. Nutr.,* 24, 1053, 1970.

300. **Dreosti, I. E., Grey, P. C., and Wilkins, P. J.,** Deoxyribonucleic acid synthesis, protein synthesis, and teratogenesis in zinc-deficient rats, *S. Afr. Med. J.,* 46, 1585, 1972.

301. **Grey, P. C. and Dreosti, J. E.,** Deoxyribonucleic acid and protein metabolism in zinc-deficient rats, *J. Comp. Pathol.,* 82, 223, 1972.

302. **Hsu, T. H. S. and Hsu, J. M.,** Zinc deficiency and epithelial wound repair: An autoradiographic study of ³H-thymidine incorporation, *Proc. Soc. Exp. Biol. Med.,* 140, 157, 1972.

303. **Sandstead, H. H., Gillespie, D. D., and Brady, R. N.,** Zinc deficiency: Effect on brain of the suckling rat, *Pediatr. Res.,* 6, 119, 1972.

304. **Dreosti, I. E. and Hurley, L. S.,** Depressed thymidine kinase activity in zinc-deficient rat embryos, *Proc. Soc. Exp. Biol. Med.,* 150, 161, 1975.

305. **Terhune, M. W. and Sandstead, H. H.,** Decreased RNA polymerase activity in mammalian zinc deficiency, *Science,* 177, 68, 1972.

306. **Scrutton, M. C., Wu, C. W., and Goldthwait, D. A.,** The presence and possible role of zinc in RNA polymerase obtained from *Escherichia coli, Proc. Natl. Acad. Sci. U.S.A.,* 68, 2497, 1971.

307. **Slater, J. P., Mildvan, A. S., and Loeb, L. A.,** Zinc in DNA polymerase, *Biochem. Biophys. Res. Commun.,* 44, 37, 1971.

308. **Springgate, C. F., Mildvan, A. S., and Loeb, L. A.,** Studies on the role of zinc in DNA polymerase, *Fed. Proc. Fed. Am. Soc. Exp. Biol.,* 32, 541, 1973.

309. **Lieberman, I., Abrams, R., Hunt, N., and Obe, P.,** Levels of enzyme activity and deoxyribonucleic acid synthesis in mammalian cells cultured from the animal, *J. Biol. Chem.,* 238, 3955, 1963.

310. **Wacker, W. E. C. and Vallee, B. L.,** Nucleic acid and metals. I. Chromium, manganese, nickel, iron, and other metals in ribonucleic acid from diverse biological sources, *J. Biol. Chem.,* 234, 3257, 1959.

311. **Tal, M.,** Metal ions and ribosomal conformation, *Biochim. Biophys. Acta,* 195, 76, 1969.

312. **Sandstead, H. H., Hollaway, W. L., and Baum, V.,** Zinc deficiency: Effect on polysomes, *Fed. Proc. Fed. Am. Soc. Exp. Biol.,* 30, 517, 1971.

313. **Weser, U., Hübner, L., and Jung, H.,** Zn^{2+}-induced stimulation of nuclear RNA synthesis in rat liver, *FEBS Lett.,* 7, 356, 1970.

314. **Scott, D. A.,** Crystalline insulin, *Biochem. J.,* 28, 1592, 1934.

315. **Hove, E., Elvehjem, C. A., and Hart, E. B.,** The physiology of zinc in the nutrition of the rat, *Am. J. Physiol.,* 119, 768, 1937.

316. **Hendricks, D. G. and Mahoney, A. W.,** Glucose tolerance in zinc-deficient rats, *J. Nutr.,* 102, 1079, 1972.

317. **Quarterman, J., Mills, C. F., and Humphries, W. R.,** The reduced secretion of and sensitivity to insulin in zinc-deficient rats, *Biochem. Biophys. Res. Commun.,* 25, 354, 1966.

318. **Boquist, L. and Lernmark, A.,** Effects of the endocrine pancreas in Chinese hamsters fed zinc-deficient diets, *Acta Pathol. Microbiol. Scand.,* 76, 215, 1969.

319. **Huber, A. M. and Gershoff, S. N.,** Effects of dietary zinc on the enzymes in the rat, *J. Nutr.,* 103, 1175, 1973.

320. **Huber, A. M. and Gershoff, S. N.,** Effect of zinc deficiency in rats on insulin release from the pancreas, *J. Nutr.,* 103, 1739, 1973.

321. **Roth, H. P., Schneider, U., and Kirchgessner, M.,** Zur Wurkung von Zinkmangel auf die Glukosetoleranz, *Arch. Tierernaehr.,* 25(8), 545, 1975.

322. **Quarterman, J. and Florence, E.,** Observations on glucose tolerance and plasma levels of free fatty acids and insulin in the zinc-deficient rat, *Br. J. Nutr.,* 28, 75, 1972.

323. **McIntyre, N., Holdsworth, C. D., and Turner, D. S.,** Intestinal factors in the control of insulin secretion, *J. Clin. Endocrinol. Metab.,* 25, 1317, 1965.

324. **Fasel, J., Hadjikhani, M. D. H., and Felder, J. P.,** The insulin secretory effect of the human duodenal mucosa, *Gastroenterology,* 59, 109, 1970.

325. **Roth, H. P. and Kirchgessner, M.,** Insulingehalte in Serum bzw. Plasma von Zinkmangelratten vor und nach glucosestimulierung, *Int. J. Vitam. Nutr. Res.,* 45, 202, 1975.

326. **Boquist, L.,** Some aspects of the blood glucose regulation and the glutathione content of the non-diabetic adult Chinese hamsters Cricetulus griseus, *Acta Soc. Med. Ups.,* 72, 358, 1967.

327. **Boquist, L.,** Alloxan administration in the Chinese hamster. I. Blood glucose variations, glucose tolerance, and light microscopical changes in pancreatic islets and other tissues, *Virchows Arch. B,* 1, 157, 1968.

328. **Engelbart, K. and Kief, H.,** Über das funktionelle Verhalten von Zink und Insulin in den β-zellen das Rattenpankreas, *Virchows Arch. B.,* 4, 294, 1970.

329. **Coombs, T. L., Grant, P. T., and Frank, B. H.,** Differences in the binding of zinc ions by insulin and proinsulin, *Biochem. J.,* 125, 62, 1971.

330. **Homan, J. D. H., Overbeek, G. A., Neutlings, J. P. J., Booiy, L. J., and Van der Vies, J.,** Corticotrophin zinc phosphate and hydroxide long acting aqueous preparations, *Lancet,* 13, 541, 1954.

331. **Cox, R. P. and Ruckenstein, A.,** Studies on the mechanism of hormonal stimulation of zinc uptake in human cell cultures: Hormone-cell interactions and characteristics of zinc accumulation, *J. Cell. Physiol.,* 77, 71, 1971.

332. **Millar, M. J., Elcoate, P. V., Fischer, M. I., and Mawson, C. A.,** Effect of testosterone and gonadotrophin injections on the sex organ development of zinc-deficient male rats, *Can. J. Biochem. Physiol.,* 38, 1457, 1960.

333. **Ku, P. K.,** Nucleic acid and protein metabolism in the zinc-deficient pig, *Diss. Abstr. Int. B,* 31, 6717, 1971.

334. **Gombe, S., Apgar, J., and Hansel, W.,** Effect of zinc deficiency and restricted food intake on plasma and pituitary LH and hypothalmic LRF in female rats, *Biol. Reprod.,* 9, 415, 1973.

335. **Lei, K. Y., Abbasi, A., and Prasad, A. S.,** Function of pituitary-gonadal axis in zinc-deficient rats, *Am. J. Physiol.,* 230, 1730, 1976.

336. **Wegener, W. S. and Romano, A. H.,** Zinc stimulation of RNA and protein synthesis in *Rhizopus nigricans, Science,* 142, 1669, 1963.

337. **Hsu, J. M., Anthony, W. L., and Buchanan, P. J.,** Incorporation of glycine-1-^{14}C into liver glutathione in zinc deficient rats, *Proc. Soc. Exp. Biol. Med.,* 127, 1048, 1968.

338. **Waters, M. D., Moore, R. D., Amato, J. J., and Houck, J. C.,** Zinc sulfate failure as an accelerator of collagen biosynthesis and fibroblast proliferation, *Proc. Soc. Exp. Biol. Med.,* 138, 373, 1971.

339. **Van Campen, D. R.,** Effects of zinc, cadmium, silver and mercury on the absorption and distribution of copper-64 in rats, *J. Nutr.,* 88, 125, 1966.

340. **Magee, A. C. and Matrone, G.,** Studies on growth, copper metabolism and iron metabolism of rats fed high levels of zinc, *J. Nutr.,* 72, 233, 1960.

341. **Hill, C. H., Matrone, G., Payne, W. L., and Barber, C. W.,** *In vivo* interactions of cadmium with copper, zinc and iron, *J. Nutr.,* 80, 227, 1963.

342. **Chvapil, M. and Walsh, D.,** A new method to control collagen cross-linking by inhibiting lysyl-oxidase with zinc, in *Proc. Workshop Conf. Connective Tissue and Aging,* Int. Congr. Ser. 264, 1972, Vogel, H. G., Ed., Excerpta Medica Foundation, Amsterdam, 1973, 226.

343. **Kulonen, E.,** Studies on experimental granuloma, in *Chemistry and Molecular Biology of the Intracellular Matrix,* Vol. 3, Balazs, E. A., Ed., Academic Press, New York, 1970, 1811.

344. **Woeesner, J. F., Jr. and Boucek, R. J.,** Connective tissue development in subcutaneously implanted polyvinyl sponge. I. Biochemical changes during development, *Arch. Biochem.,* 93, 85, 1961.

345. **Hsu, J. M.,** Zinc as related to cystine metabolism, in *Trace Elements in Human Health and Disease,* Vol. 1, Prasad, A. S., Ed., Academic Press, New York, 1976, 295.

346. **Somers, M. and Underwood, E. J.,** Studies of zinc nutrition in sheep. II. The influence of zinc deficiency in ram lambs upon the digestibility of the dry matter and the utilization of the nitrogen sulphur of the diet, *Aust. J. Agric. Res.,* 20, 899, 1969.

347. **Warren, L., Glick, M., and Nass, M.,** Membranes of animal cells. I. Methods of isolation of the surface membrane, *J. Cell. Physiol.,* 68, 269, 1966.

348. **Nicholson, V. J. and Veldstra, H.,** The influence of various cations on the binding of colchicine by rat brain homogenates: stabilization of intact neurotubules by zinc and calcium ions, *FEBS Lett.,* 23, 309, 1972.

349. **Karl, L., Chvapil, M., and Zukoski, C. F.,** Effect of zinc on the viability and phagocytic capacity of peritoneal macrophages, *Proc. Soc. Exp. Biol. Med.,* 142, 1123, 1973.

350. **Mustafa, M. G., Cross, C. E., and Hardie, J. A.,** Localization of Na^+-K^+, Mg^+ adenosinetriphosphatase activity in pulmonary alveolar macrophage sub-cellular fractions, *Life Sci.,* 9, 947, 1970.

351. **Mustafa, M. G., Cross, C. E., Munn, R. J., and Hardie, J. A.,** Effects of divalent metal ions on alveolar macrophage membrane adenosine triphosphatase activity, *J. Lab. Clin. Med.,* 77, 563, 1971.

352. **Donaldson, J., St. Pierre, T., Minnich, J., and Barbeau, A.,** Seizures in rats associated with divalent cation inhibition of Na^+-K^+-ATPase, *Can. J. Biochem.,* 49, 1217, 1971.

353. **Robinson, J. D.,** Divalent cations as allosteric modifiers of the $(Na^+ + K^+)$-dependent ATPase, *Biochim. Biophys. Acta,* 226, 97, 1972.

354. **Romeo, D., Zabucchi, G., Soranzo, M. R., and Rossi, F.,** Macrophage metabolism: activation of NADPH oxidation by phagocytosis, *Biochem. Biophys. Res. Commun.,* 45, 1056, 1971.

355. **Sbarra, A. J., Paul, B. B., Jacobs, A. A., Straus, R. R., and Mitchell, G. W., Jr.,** Biochemical aspects of phagocytic cells as related to bactericidal function, *J. Reticuloendothel. Soc.,* 11, 492, 1972.

356. **Chvapil, M.,** New aspects in the biological role of zinc: a stabilizer of macromolecules and biological membranes, *Life Sci.,* 13, 1041, 1973.

357. **Chvapil, M., Zukoski, C. F., Hattler, B. G., Stankova, L., Montgomery, D., Carlson, E. C., and Ludwig, J. C.,** Zinc and cells, in *Trace Elements in Human Health and Disease,* Vol. 1, Prasad, A. S., Ed., Academic Press, New York, 1976, 269.

358. **Sacchetti, G., Gibelli, A., Bellani, D., and Montanari, C.,** Effect of manganese ions on human platelet aggregation *in vitro, Experientia,* 30, 374, 1974.

359. **Ruhl, H., Kirchner, H., and Bochert, G.,** Kinetics of the Zn^{2+} stimulation of human peripheral lymphocytes *in vitro, Proc. Soc. Exp. Biol. Med.,* 137, 1089, 1971.

360. **Chesters, J. K.,** The role of zinc ions in the transformation of lymphocytes of phytohaemaglutinin, *Biochem. J.,* 130, 133, 1972.

361. **Alford, R. H.,** Metal cation requirements for phytohemaglutinin-induced transformation of human peripheral blood lymphocytes, *J. Immunol.,* 104, 698, 1970.

362. **Berger, N. A. and Skinner, A. M.,** Characteristics of lymphocyte transformation induced by zinc ions, *J. Cell Biol.,* 61, 45, 1974.

363. **Kroneman, J., v.d. Mey, G. J. W., and Helder, A.,** Hereditary zinc deficiency of Dutch friesian cattle, *Zentralbl. Veterinaermed. Reihe A,* 22, 201, 1975.

363a. **Frost, P., Chen, J. C., Amjad, H., and Prasad, A. S.,** The "null lymphoid" cell in sickle cell disease, *Clin. Res.,* 240, 570A, 1976.

363b. **Frost, P., Chen, J. C., Rabbani, P., Smith, J., and Prasad, A. S.,** The effect of zinc deficiency on the immune response, in *Zinc Metabolism: Current Aspects in Health and Disease,* Brewer, G. J. and Prasad, A. S., Eds., A. R. Liss, New York, in press.

364. **Brewer, G. J. and Oelshlegel, F. J., Jr.,** Antisickling effects of zinc, *Biochem. Biophys. Res. Commun.,* 58, 854, 1974.

365. **Dash, S., Brewer, G. J., and Oelshlegel, F. J., Jr.,** Effect of zinc on haemoglobin binding by red blood cell membranes, *Nature,* 250, 251, 1974.

366. **Hoffman, J. F.,** Physiological characteristics of human red blood cell ghosts, *J. Gen. Physiol.,* 42, 9, 1958.

367. **Weed, R. I., LaCelle, P. L., and Merrile, E. W.,** Metabolic dependence of red cell deformability, *J. Clin. Invest.,* 48, 795, 1969.

368. **Palek, J., Curby, W. A., and Lionetti, F. J.,** Effects of calcium and adenosine triphosphate on volume of human red cell ghosts, *Am. J. Physiol.,* 220, 19, 1971.

369. **Eaton, J. W., Skelton, T. D., Swofford, H. A., Kolpin, C. E., and Jacob, H. A.,** Elevated erythrocyte calcium in sickle cell disease, *Nature,* 246, 105, 1973.

370. **Vallee, B. L and Wacker, W. E. C.,** The metalloproteins, in *The Proteins,* Vol. 5, Neurath, H., Ed., Academic Press, New York, 1972, 143.

371. **Korant, B. D., Lonberq-Holm, K. K., Noble, J., and Stasny, J. T.,** Naturally occurring and artificially produced components of three rhinoviruses, *Virology,* 48, 71, 1972.

372. **Korant, B. D., Kauer, J. C., and Butterworth, B. E.,** Zinc ions inhibit replication of rhinoviruses, *Nature,* 248, 588, 1974.

373. **Butterworth, B. E. and Korant, B. D.,** Characterization of the large picornaviral polypeptides produced in the presence of zinc ion, *J. Virol.,* 14, 282, 1974.

374. **Papp, J. P.,** Metal fume fever, *Postgrad. Med.,* 43, 160, 1968.

375. **Murphy, J. V.,** Intoxication following ingestion of elemental zinc, *JAMA,* 212, 2119, 1970.

376. **Gallery, E. D. M., Blomfield, J., and Dixon, S. R.,** Acute zinc toxicity in hemodialysis, *Br. Med. J.,* 4, 331, 1972.

377. **Heller, V. G. and Burke, A. D.,** Toxicity of zinc, *J. Biol. Chem.,* 74, 85, 1927.

378. **Furchner, J. E. and Richmond, C. R.,** Effect of dietary zinc on the absorption of orally administered Zn^{65} *Health Phys.,* 8, 327, 1968.

379. *The Merck Index,* 7th Ed., Merck, Rahway, N.J., 1960, 1118.

380. **Csata, S., Gallays, F., and Toth, M.,** Akute Niereninsuffizienz als Folge einer Zinkchloridvergiftung, *Z. Urol.,* 61, 327, 1968.

381. **Osol, A., Farrar, G. E., Jr., and Pratt, R.,** *Dispensatory of the U.S.,* 25th ed., J. B. Lippincott, Philadelphia, 1955, 1520.

382. **Van Reen, R.,** Zinc toxicity in man and experimental species, in *Zinc Metabolism,* Prasad, A. S., Ed., Charles C Thomas, Springfield, Ill., 1966, 411.

383. **Kagi, J. H. R. and Vallee, B. L.,** Metallothionein: a cadmium and zinc-containing protein from equine renal cortex, *J. Biol. Chem.,* 235, 3460, 1960.

384. **Kagi, J. H. R. and Vallee, B. L.,** Metallothionein: a cadmium and zinc-containing protein from equine renal cortex, *J. Bio. Chem.,* 236, 2435, 1961.

385. **Richards, M. P. and Cousins, R. J.,** Mammalian zinc homeostasis: requirement for RNA and metallothionein synthesis, *Biochem. Biophys. Res. Commun.,* 64, 1215, 1975.

386. **Squibb, K. S. and Cousins, R. J.,** Synthesis of metallothionein in a polysomal cell-free system, *Biochem. Biophys. Res. Commun.,* 75, 806, 1977.

INDEX

A

Acrodermatitis enteropthica as cause of zinc deficiency, 16, 29
Alcohol dehydrogenase
   activity in tissue of zinc-deficient animals, 46, 47, 48
   effect of zinc deficiency on, 45
Alcohol, effect on zinc excretion, 17
Aldolase, activity in zinc deficiency, 49
Alkaline phosphatase
   activity in tissues of zinc-deficient animals, 21
   changes in, in zinc deficiency, 4, 5, 10, 17
   effect on of zinc deficiency, 41—43
   hemoglobin retention, effect of zinc on, 66
Alopecia in zinc deficiency, 1
Amino acid concentrations in serum, 34
Anemia
   in iron deficiency, 2, 4
   liver function studies of patients, 7
   sickle cell, see Sickle cell disease, 16
Animals, zinc deficiency in, 1
Anorexia in zinc deficiency, 14, 16
Antianabolic drugs, zinc deficiency resulting from use, 23
Antimetabolites, zinc deficiency resulting from use of, 23
Aspergillus niger, requirement of zinc for growth, 1

B

Bedsores, zinc treatment of, 23
Birds, zinc requirement, 1
Blood, zinc content, 30
Bone, zinc concentrations in, 30
Burns as cause of zinc deficiency, 16, 21

C

Calcium
   incorporation into sickled cells, effect of zinc on, 66
   levels, relationship to zinc concentrations, 21
Calves, zinc deficiency in, 1
Cancer as cause of zinc deficiency, 16, 21
Carbonic anhydrase, effect of zinc deficiency on, 44
Carboxypeptidase
   activities in tissues of zinc-deficient animals, 46, 47, 48
   effect of zinc deficiency on, 43, 44
Celiac disease, zinc deficiency in patients, 12
Cell membranes, effect of zinc on, 63—65
Chelating agents, effect on zinc absorption, 34
Children, zinc deficiency in, 11, 12, 13
China, studies of dwarfism and hypogonadism in, 8
Chronic infections as cause of zinc deficiency, 16, 21
Cirrhosis of the liver as cause of zinc deficiency, 16—20, 22
Clay eating, see Geophagia
Colitis, ulcerative, zinc concentrations in, 20
Collagen

content in sponge connective tissue, 60
   diseases, zinc deficiency in, 23
   metabolism, role of zinc in, 57—63
   sedimentation patterns in sponge connective tissue, 62
Conjunctivitis, in zinc deficiency, 1
Crohn's disease, zinc concentrations in, 20
Cystic fibrosis, zinc deficiency in patients, 12, 22
Cystine metabolism, role of zinc in, 63

D

Denver, children in, zinc deficiency in, 13
Diabetes, zinc deficiency in, 23
Diet
   as cause of zinc deficiency, 16, 17, 21
   zinc in, absorption, 32, 33, 35, 37, 38
Digits, deformation, in zinc deficiency, 1
Disaccharidase deficiency, zinc deficiency in patients, 12
DNA
   concentrations, effect of zinc deficiency on, 54, 59
   content in zinc-deficient animals, 48, 49
   effect on of zinc deficiency, 51, 52
   role of zinc in metabolism, 41
   role of zinc in polynucleotide conformation, 53
DNA polymerase
   effect of zinc deficiency on, 51
   zinc as constituent, 40, 41
Dogs, zinc deficiency in, 1
Dwarfism in zinc deficiency, 2—12

E

Egypt, zinc deficiency studies in, 4—12
Elephants, iron deficiency in, 2
Emaciation in zinc deficiency, 1
Emesis, in zinc deficiency, 1
Enteritis, zinc deficiency in patients, 12, 19
Enzymes
   with zinc constituent, see Zinc metalloenzymes
   zinc-independent, activities in zinc deficiency, 49
Esophagus
   effect on of iron deficiency, 4
   zinc concentrations in, 30
Eye, zinc concentrations in, 30

F

Food additives, effect on zinc absorption, 17

G

Gastrectomy as cause of zinc deficiency, 16, 20
Gastrointestinal disorders, zinc deficiency in, 19, 20
Genetic causes of zinc deficiency, 16, 24—29

Geophagia
  as cause of zinc deficiency, 16, 17
  association with nutritional deficiency, 2, 4, 6
  clay as source of zinc, 34
Glucose tolerance in zinc deficiency, 53—56
Glutamic dehydrogenase, effect of zinc deficiency on, 45
Growth
  in patients receiving supplemental zinc, 9, 11
  of zinc-deficient Denver children, 13
  retarded
    idiopathic, zinc concentrations in, 20
    in acrodermatitis enteropathica, 29
  zinc deficiency affecting
    in animals, 1, 4
    in humans, 2—15
Growth hormones, role of zinc in function, 57

H

Hair, zinc concentrations in, 5, 6, 9, 12—16
Heart, zinc concentration, 30
Hemodialysis, plasma zinc concentrations in, 22
Hemoglobin retention by red cells, effect of zinc on, 66
Hens, zinc deficiency in, 1
Hepatosplenomegaly
  in nutritional deficiency, 2, 4
  zinc treatment, 9
Hookworm infection in nutritional deficiency patients, 5, 6, 22
Hormones, role of zinc in function, 53—57
Hospital diet, zinc content, 37
Hyperkeratosis in zinc deficiency, 1
Hyperzincemia, familial, 29
Hyperzincuria
  alcohol as cause, 17
  in cirrhosis of the liver, 17, 20
  in sickle cell disease, 17, 20
Hypogeusia in zinc deficiency, 14
Hypogonadism
  in acrodermatitis enteropathica
  in nutritional deficiency, 2, 6—10
  in sickle cell disease, 24
Hypopharynx, effect on of iron deficiency, 4
Hypoplasia of thymus in cattle, zinc deficiency in, 29

I

Iatrogenic causes of zinc deficiency, 16, 23
Immune responses, role of zinc in, 65
Indolent ulcer, plasma zinc concentrations in, 22
Infants, zinc deficiency in, 14
Infection, zinc deficiency in, 23
Injury, zinc deficiency in, 23
Insulin
  function in glucose levels, 53
  secretion, effect of zinc deficiency on, 56
  zinc content, 53
Intestinal mucosa disease as cause of zinc deficiency, 16

Intestine, zinc absorption from, 32, 33
Intravenous therapy, effect on zinc concentrations, 23
Iron, zinc deficiency studies in, 2—11
  effects of, 2
Isocitric dehydrogenase, activity in tissues of zinc-deficient animals, 46, 47

J

Japanese quail, zinc deficiency in, 1

K

Keratitis in zinc deficiency, 1
Kidney dialysis
  as cause of zinc deficiency, 16
  effect of diet on zinc levels, 21
Kidney disease as cause of zinc deficiency, 16, 20, 21
Kidney, zinc concentrations in, 30
Koilonychia in iron deficiency, 4
Kwashiorkor, plasma zinc levels in patients, 12, 13

L

Lactic dehydrogenase
  activity in tissues of zinc-deficient animals, 46, 47, 48, 49
  effect of zinc deificiency on, 45
Leg ulcers, zinc treatment of, 21, 23
Liver
  cirrhosis of, see Cirrhosis of the liver
  zinc concentrations in, 30
Liver function test, use in zinc deficiency studies, 5, 7
Lung, zinc concentrations in, 30
Lupus erythematosus, zinc deficiency in, 23
Lymphocytes, role of zinc in DNA synthesis, 65

M

Malic dehydrogenase, effect of zinc deficiency on, 45
Malnutrition, generalized, zinc deficiency associated with 12
Metabolism of zinc
  process, 30—39
  relationship to malignancy, 41
Metallothionein, role of zinc in synthesis, 67
Metal fume fever, 67
Mice, zinc deficiency in, 1
Mongolism, plasma zinc concentrations in, 22, 29
Morocco, zinc-deficiency dwarfism in, 11
Muscle, zinc concentrations in, 30
Myocardial infarction, plasma zinc concentrations in, 22

N

NADH diaphorase, levels during zinc deficiency, 49

Neoplastic disease as cause of zinc deficiency, 16, 21
Nephrotic syndrome as cause of zinc deficiency, 16
Nutrition, see Diet

O

Oral contraceptives, plasma zinc decrease following use, 24
Ossification, retarded, in zinc deficiency, 1

P

Pancreas, zinc concentrations in, 30
Pancreatic insufficiency as cause of zinc deficiency, 16
Parasitic infections
    as cause of zinc deficiency, 16, 22
    associated with nutritional deficiencies, 5, 6, 8
Penicillamine therapy as cause of zinc deficiency, 16, 23
Pica, see also Geophagy
    in zinc deficiency, 16
Pigs, see Swine
Plasma, zinc concentrations in, 5, 6, 8, 13—16, 22, 30
Portugal, zinc-deficiency dwarfism in, 11
Pregnancy, zinc deficiency in, 16, 22, 24
Prostate, zinc concentrations in, 30
Protein content of sponge connective tissue, 60
Psoriasis as cause of zinc deficiency, 16, 21
Pulmonary infection, plasma zinc concentrations in, 22
Pyridoxal phosphokinase, activity during zinc deficiency, 49

R

Rats
    growth, zinc required for, 1
    iron deficiency in, 2
Red cells, see Blood
Renal dialysis
    as cause of zinc deficiency, 16
    effect of diet on zinc levels, 21
Renal disease as cause of zinc deficiency, 16, 20, 21
Rheumatoid arthritis, zinc deficiency in, 23, 30
RNA
    concentrations, effect of zinc deficiency on, 54, 59
    content in zinc-deficient animals, 48, 49
    effect on zinc deficiency, 51, 52
    role of zinc in metabolism, 41
    role of zinc in polynucleotide conformation, 53
RNA polymerase
    effect of zinc deficiency on, 51
    zinc as constituent, 40, 41

S

Schistosomiasis in nutritional deficiency patients, 5, 6, 22
Serum zinc content in relation to that of plasma, 30
Serum zinc levels in Iranian children, 12

Sex hormones, role of zinc in function, 57
Sexual development, see also Hypogonadism
    retarded, in nutritional deficiency, 2, 6—10
Sickle cell disease
    as cause of zinc deficiency, 16, 17, 20, 24—28
    decreased enzyme activity in, 45
    effect of zinc on cell filterability, 66, 67
    hormone levels in, 25
    symptoms, 24
    weight change during zinc therapy, 28
    zinc contents of erythrocytes in patients, 27
    zinc excretion in, 26
    zinc treatment, 25, 26
Skin
    disorders, as cause of zinc deficiency, 16, 21
    hyperkeratinization, in zinc deficiency, 1
    lesions, in zinc deficiency, 1
Sponge connective tissue
    collagen content in zinc deficiency, 60
    effect of zinc deficiency on, 54
    polyribosomes, effect on of zinc deficiency, 61
    sedimentation patterns of collagen, 62
Sprue, zinc concentrations in, 20
Starch, laundry, as cause of zinc deficiency, 16, 17
Starvation, excretion of zinc during, 38, 39
Steatorrhea as cause of zinc deficiency, 16, 19
Succinic dehydrogenase, activity in tissue of zinc-deficient animals, 46—48
Surgical trauma as cause of zinc deficiency, 16
Swine
    enzyme changes in zinc deficienty, 4
    parakeratosis in, zinc as cure for, 1

T

Taste impairment in zinc deficiency, 14
Testicles, see also Hypogonadism
    atrophy of, in zinc deficiency, 1, 4
    zinc concentrations in, 30
Thymidine kinase
    as zinc-dependent enzyme, 41
    effect of zinc deficiency on, 51, 52
Toxicity of zinc, 67
Tuberculosis, plasma zinc concentrations in, 22
Turkey, zinc-deficiency dwarfism in, 11

U

Uremia, plasma zinc concentrations in, 22
Urine, zinc excretion, 6, 8

V

Viruses, effect of zinc on, 67
Vitamin A levels in zinc deficiency, 14

W

Weight
  effect on of zinc deficiency, 54, 59
  reduction, zinc balance during, 39
Wounds, zinc treatment of, 21

Z

Zinc absorption process, 32—35
Zinc deficiency
  causes of, 16
  diagnosis, 16
  dwarfism in, 2—12
  effect on cystine metabolism, 63
  effect on DNA and RNA, 51
  effect on DNA and RNA concentrations, 54, 59
  effect on enzyme activity, 41—53
  effect on glucose tolerance, 53—56
  effect on protein concentrations, 54
  effect on sponge connective tissue, 54, 60, 61
  effect on weight, 54, 59
  impaired protein synthesis in, 51
  iatrogenic causes, 16, 23
  in acrodermatitis enteropethica, 29
  in animals, 1
  in burns, 16, 21
  in cirrhosis of the liver, 16, 17, 22
  in collagen diseases, 23
  in cystic fibrosis, 22
  in Denver children, 13, 14
  in diabetes, 23
  in enteritis, 19
  in gastrointestinal disorders, 19
  in genetic disorders, 16, 29
  in hemodialysis, 22
  in indolent ulcer, 22
  in infants, 14
  in mongolism, 22
  in myocardial infarction, 22
  in neoplastic disease, 16, 21
  in parasitic infections, 16, 22
  in preadolescent children, 11
  in pregnancy, 16, 22, 24
  in pulmonary infection, 22
  in renal disease, 16, 20
  in sickle cell disease, 24—29
  in skin disorders, 16, 21
  in steatorrhea, 19
  in tuberculosis, 22
  in uremia, 22
  in various populations, 11
  in deficiency associated with, 4
  male-female susceptibility, 9
  nutritional factors, 16, 17
  oral contraceptives affecting, 24
  role in infections, 65
  secondary, in certain diseases, 12
  signs of
    in animals, 1, 4
    in humans, 2—15
  total protein in, 50
  urea synthesis affected by, 19
Zinc metalloenzymes
  definition, 41
  effect of zinc deficiency on, 41—53
  function of zinc in enzymes, 41
  sources of, 40
Zinc treatment
  effect on urinary zinc excretion, 38
  of bedsores, 23
  of leg ulcers, 21, 23
  of rheumatoid arthritis, 30
  of sickle cell disease, 25, 26, 28
  of wounds, 21
  possible toxic effects, 68